ESTABLISHMENT OF FIRST GENE SANCTUARY IN INDIA FOR CITRUS IN GARO HILLS

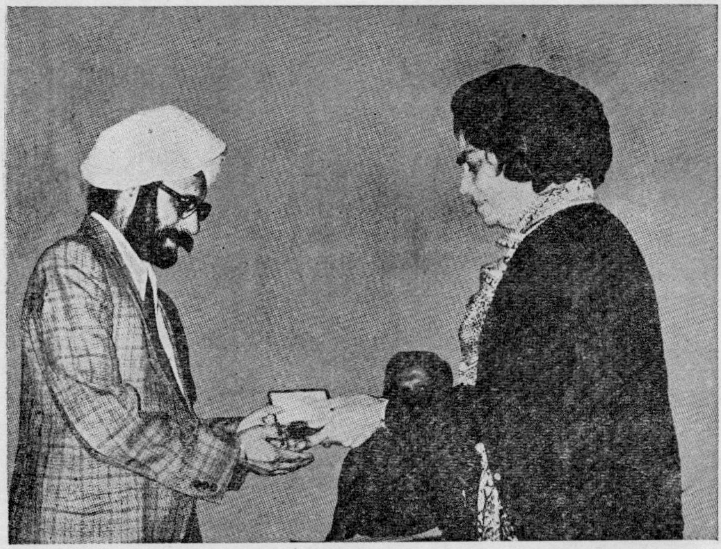

The author is seen receiving the Fakhruddin Ali Ahmad Award for Agricultural Research in Tribal India from Begum Abida Ahmad.

ESTABLISHMENT OF FIRST GENE SANCTUARY IN INDIA FOR CITRUS IN GARO HILLS

BHAG SINGH
National Bureau of Plant Genetic Resources

Foreword by

M.S. SWAMINATHAN, F.R.S.

CONCEPT PUBLISHING COMPANY
NEW DELHI

The National Bureau of Plant Genetic Resources does not hold responsibility for the opinions expressed in this book; such responsibility rests solely with the author.

First Published 1981

Bhag Singh (*b*. 1935-)

Published by
Naurang Rai
Concept Publishing Company
H-13, Bali Nagar
NEW DELHI-110015 (India)

Printed by
Patel Enterprises
at Sunil Printers,
1067, Ajay Palace, Naraina
New Delhi-110028

Foreword

GENE segregation, recombination and mutation provide opportunities for the occurrence of extensive genetic variability in all living organisms. This forms the basis for the popular statement "variety is the spice of life". Depending upon environmental and biotic factors, different genotypes survive under the most hospitable conditions. Human selection and interference can change the situation and to-day we find that many plants and animals thrive well under conditions which are totally different from those occurring in their original habitats. The frontiers of adaptation have been extended by man. At the same time, human interference with the habitat requirements of different species has also accelerated the pace of species extinction. O.H. Frankel and M.E. Soule in a recent book on *Conservation and Evolution* published by the Cambridge University Press, have stated: "Throughout the history of life, there has never been as wanton nor as rapid an agent of habitat destruction as Twentieth Century man". Every species being only as safe as its habitat, there is a grave danger of our losing valuable genetic material if steps are not taken to protect the natural habitats of wild species or make alternative arrangements to conserve natural variability. The rate of destruction of tropical forests to-day has been estimated at about 47 hectares per minute. At this rate the tropical forests will become historical curiosities within a century.

Frankel and Soule have classified the various factors which contribute to species extinction as follows :

1. Biotic factors
   (a) competition
   (b) predation
   (c) parasitism and disease
2. Isolation

3. Habitat alteration
   - (a) slow geological change
   - (b) climate
   - (c) catastrophe
   - (d) man

In recent years, organised arrangements for the collection and conservation of crop genetic resources have been streamlined. The International Board for Plant Genetic Resources (IBPGR) organised by the Consultative Group on International Agricultural Research in 1974 has been successful in stimulating and co-ordinating international concern for crop genetic resources. Presently the IBPGR is dealing with 120 species of food plants and other plants of global or regional importance. At a recent meeting the following priority areas have been identified in relation to the programmes of IBPGR :

— additional cold stores for seed conservation should be provided to strengthen the international network ;

— the use of tissue culture techniques should be expedited for conservation of clonal materials, probably guided in the first phase by an IBPGR expert working group;

— work on the characterization and evaluation of germplasm in genebanks should be speeded up and the findings transmitted to potential users ;

— basic informational data should always be sent out with samples.

The conservation programmes of IBPGR as well as the National Bureau of Plant Genetic Resources of our country have largely centred around annual crop plants. Fruit and forest trees are yet to receive similar attention. In many cases *in situ* conservation is the best method of preserving the variability in the wild species of potential value to man. This kind of conservation is akin to the conservation of natural ecosystems. In eco-system conservation, the emphasis is on representativeness of particular communities within the ecosystem as a whole. In long-term reserves such as the biosphere reserves, the time scale of concern is infinity. Attempts are now being made particularly in the United States and the

USSR to protect the endangered plant species which occur usually in restricted sites. There is, however, need for the more widespread establishment of genetic reserves or gene sanctuaries in which wild relatives of crop species can be protected and the population biology of individual species can be studied. The studies of Dr. Bhag Singh, which form the basis of this book, relate to the preservation of the variability in *Citrus* in the north-eastern Himalayan region. Dr. Bhag Singh has in particular dealt with in detail the potential for conserving the genetic wealth of *Citrus indica* in the Tura range of the Garo Hills in Meghalaya. We are fortunate in our citrus wealth. The north-eastern Himalayan region has also proved to be a veritable mine of valuable genes in many other crops including rice and maize. Unfortunately due to the practice of shifting cultivation or *jhuming* and widespread habitat destruction, much of this genetic wealth faces the threat of extinction. We, therefore, owe a debt of gratitude to Dr. Bhag Singh and the National Bureau of Plant Genetic Resources of the Indian Council of Agricultural Research for their timely initiative in surveying the habitat of *Citrus indica* and suggesting a plan for conserving this unique genetic wealth for posterity.

Yojana Bhawan
New Delhi-110001
14 October 1981

M.S. SWAMINATHAN
Member
Planning Commission

Preface

THE INFORMATION given in this publication was gathered at the regional station of the National Bureau of Plant Genetic Resources, at Shillong, financed by the Indian Council of Agricultural Research. The investigations pertaining to the establishment of a Citrus Gene Sanctuary in Meghalaya were initiated in 1978 at Shillong.

The investigations summarised in this book pertain to the preservation of the native habitat of the wild orange (*Citrus indica*) a progenitor of the mandarin orange. It is hoped that this book will help to stimulate interest among the youth in the genetic endowments of our country.

This project was conceived by Dr. M.S. Swaminathan, F.R.S., former Director General of the Indian Council of Agricultural Research and Secretary, Department of Agricultural Research and Education, presently Member, Planning Commission and Chairman, Science Advisory Committee to the Cabinet. He used the term "gene sanctuary" for stressing the need for preserving the environmental conditions under which certain specific genotypes flourish. His efforts made it possible to identify the wild orange faced with danger of extinction in this part of the world. This concept was vigorously implemented with his collaboration.

My wife Smt. Balwant Kaur and my son Shri Irvinder Singh left no stone unturned to familiarize themselves with the tribals which laid the foundation to initiate the work of these systematic exploratory surveys. Their active participation and encouragement were my main source of inspiration. My wife could not accompany me in the later part of the project due to ill-health.

A committee of experts headed by Dr. B. P. Pal, F.R.S. Chairman, National Committee on Environmental Planning and Coordination, approved the proposal based on exploratory and ecological surveys for the establishment of a Citrus Gene Sanctuary in Meghalaya. Subsequently, in 1980, the ICAR constituted a Task Force under the Chairmanship of Shri J.C. Nampui, I.A.S., Chief Secretary, Government of Meghalaya to work out an operational blue print for the establishment and development of this sanctuary. Recently, the blue print of the site has been prepared.

Many people have helped in various ways, to all of whom I should like to express my sincere thanks. The work would not have been possible but for the keen interest, able guidance and constant encouragement of Dr. O.P. Gautam, Director-General ICAR and Secretary, Department of Agricultural Research and Education. I am also equally indebted to Dr. K.L. Mehra, Director, National Bureau of Plant Genetic Resources for his keen interest, able guidance, constant encouragement and for providing all the facilities during the research work.

I wish to express my gratitude to the officials of the Government of Meghalaya for their painstaking efforts towards the establishment of the First Gene Sanctuary in India for Citrus in Garo Hills. I wish to express my gratitude for their cooperation, generous help, guidance and valuable suggestions during the present investigations. My special thanks are due to Shri L.P. Singh, former Chairman, North Eastern Council and Governor, and Chief Minister, Meghalaya for their support and encouragement.

I am thankful to Dr. B.P. Pal, F.R.S., former Chairman NCEPC; Dr. D.N. Borthakur, Director, ICAR Research Complex for North Eastern Region; Dr. S.Z. Quasim, Secretary, Department of Environment; Dr. H. K. Jain, Director, Indian Agricultural Research; Dr. J. Joseph, Deputy Director, Botanical Survey of India; Dr. P.S. Ramakrishnan, Professor of Botany, North Eastern Hill University for help, guidance and valuable suggestions. I am deeply indebted to Prof. M.G.K. Menon, I.A.S., Secretary, Science and Technology for valuable suggestions and to Shri N.D. Jayal, I.A.S., Joint Secretary, Department of Environment for his keen interest, valuable suggestions and guidance since the inception of these studies. I am also thankful to the other staff of Science and Technology for their helpful suggestions.

I wish to express my deep appreciation for the valuable help and cooperation extended by the staff members of National Bureau of Plant Genetic Resources. I wish to thank especially Prof. M.W. Hardas and Dr. R.K. Arora.

I am grateful to Col. S.M. Chadha, Director and Shri K. Baburajan, Survey of India and other staff members for their help and cooperation.

I am thankful to Dr. R.R. Misra, Prof. of Botany; Dr. Asket Singh, Deputy Director, Zoological Survey of India for their help. My sincere thanks are also due to Dr. S. K. Jain, Director and staff members of Botanical Survey of India, Howrah, Shillong and Dehra Dun and Forest Research Institute, Dehra Dun for their help and for providing facilities to use their herbaria.

Among those to whom I owe indebtedness for constant encouragement and inspiration, the name of Smt. Mina Swaminathan deserves special mention.

Finally, the National Bureau of Plant Genetic Resources does not hold responsibility for the opinions expressed in this book, such responsibility rests with the author alone.

BHAG SINGH

Contents

List of Illustrations

List of Figures

List of Tables

1

Introduction

SOME believe man as a force above nature. The Creator required man's active and rational participation for His plans. Although man is selfish and shortsighted, he is not irrational. His self destruction was no more ordained than his felicity was supposed to have been. If he has the capability to ruin nature, he might also mend it. It is true that the welfare of future generations matters more than any immediate considerations. Anyone wielding a hoe or an axe knows what he is doing but more important is it to assess the cumulative effect of all axes and hoes. Man unequivocally depends on nature which includes soil, water, plants and animals. He unwittingly destroys the supporting fabric of nature in obtaining them. It is therefore essential to understand the effects of environment. This will help to restore and maintain the natural environment as long as man tenants the earth. Civilized man has committed ravages and destroyed the balance which nature has established between organic life and inorganic matter. The earth was given to man for usufruct alone, not for consumption, still less for profligate waste. Nature has left it within the power of man irreparably to derange the combinations of organic life and the inorganic world. The organic and inorganic matter are bound together by mutual relations and adaptations in a long continuum of the established conditions of each at any given time and place.

The geological causes whose actions cannot be guided are partially responsible for the decay of flourishing countries. But the direct violence of hostile human force has the manifold role to ruin the fairest and most fruitful nature. This could be due to man's ignorant disregard of the sophisticated laws of nature or may be the incidental consequences of war and of civil tyranny and misrule. It is but recently that man has awakened and felt the necessity of restoring the disturbed harmonies of nature, whose well-balanced influence is so rewarding to all her organic world. Man is a uniquely potent creature of the ecosystem hence continued efforts should be made to provide him and his successors with the means to live and to improve their lot upon this earth. The conscious changes brought about by the races that inhabit the regions of the world are the products of complicated forces which not only alarm the physical balances but organic creation also. While every effort is being made to increase productivity one should not ignore to conserve and enhance the quality of the ecosystem.

Haphazard systems in agriculture are often found in many parts of the world. Historically, agriculture and forestry are looked upon as competitive and not complementary activities. It is conceded that forests were cleared to make room for agriculture, the world over. The farmers took the trees as their enemies and preferred to grow their crops in isolation with potential competitors ruthlessly removed. The advanced countries and the developing countries including India have suffered from ecological as well environmental damage resulted from expanding civilization. Every step taken by mankind towards progress has only increased its dependence on nature. The advancement in civilization coupled with increased population, by and large, has definitely threatened the life supporting systems. Forests do not serve the needs only of the local communities but also people far beyond. Developing countries like India have suffered irrevocable damage by destroying innumerable diversified trees in the quest of food, fuel and fodder. The North Eastern Himalayan region comprising the states of Arunachal Pradesh, Assam, Manipur, Meghalaya, Mizoram, Nagaland, Tripura and Sikkim is not an exception. This region is primarily inhabited by the tribals practising the age-old methods of 'Shift cultivation' or 'Jhuming'. In this method people find it absolutely necessary to clear the forests to provide food for their very existence and obviously this causes the mountain ecosystem to be denuded of natural vegetation. The clearing of the forests is made during the winter months and burning starts with the summer, for a year or so, then the yield of the crops slows down and a new forest is brought under this practice. The tribals are considered the custodians for the fascinating genetic wealth, since the north-eastern region is primarily inhabited by them. These tribals are yet in the wake of development which obviously also posed the need for clearing forests for the construction of a communication network and an increased demand for construction purposes.

There has been a great concern recently to locate and preserve the wild relatives of cultivated plants. The wild sugarcane variety, *Saccharum spontaneum* proved to be the most useful stock in saving the sugar industry from disaster caused by a virus disease. Vavilov (1950) in his note on the "The phytogeographic basis of plant breeding" has indicated "It is natural that the first step in breeding should be the maximum utilization of local material, the segregation of their most productive and valuable forms. The success of breeding in our country (Russia) in recent years has been chiefly based on this segregation of the most valuable forms from local population. Our best varieties of winter and spring wheat, rye, flax and barley produced in recent years by our breeding stations, are chiefly the results of selections from local varieties". About 60 years ago N.I. Vavilov demonstrated the scientific and economic significance of conservation of plant genetic resources. Subsequently, it tended to be more widespread. Consequently, an International Board for Plant Genetic Resources is now functioning under the aegis of the Consultative Group on International Agricultural Research. The IBPGR in a systematic, scientific and sustained manner has led to the identification of priorities and action programmes in the collection, evaluation and conservation of plant genetic wealth.

India is endowed with immense agri-horticultural crops. The natural diversity of many crops prevails in the forests i.e. for *Citrus*, *Musa* and *Mangifera* in the north-eastern Himalayan region. It was proposed to plan systematic studies for preserving the pockets of such habitats where considerable genetic variability occurs in economically important plants. It was urged to preserve the important materials by way of gene sanctuaries. The gene sanctuary is a mechanism to preserve the genetic diversity of endangered species by protecting the ecosystem in which it occurs naturally.

The citrus fruits which include oranges, limes, lemons citrons rank third in acreage among the sub-tropical fruits of the world. The orange, lemon and lime are known for their taste and daily domestic use. The loose skinned oranges with known speciality are called mandarin oranges. Because of deforestation, several natural growing species are facing the danger of extinction. Nature is the best conservatory, but due to man's encroachment it is dwindling fast. Consistent efforts are being made both at national and international levels to identify and to preserve such useful variability. It is well recognized that the judicious use of the natural resources alone can save mankind from unforeseeable disasters.

There has been an increasing stress in the conservation area where systematic scientific efforts will have to be supported by social and political consciousness and action. A classical example of such a conservation has been set at Leningrad during World War II. The Vavilov All Union Institute of Plant Industry which had a rich collection of germ plasm of wheat potato etc. continued to grow during the war in order to preserve them. Seemingly, these were the only crops growing in the area wherein the citizens were suffering from acute hunger. A large number died due to starvation. It is to their great credit that neither potatoes nor grain were stolen by anyone from the fields or gene bank of the Vavilov Institute of Plant Industry. This is the richest tribute which one could possibly render to the work of Vavilov and to the spirit of conservation. It is precisely this type of social commitment to the conservation of genetic wealth that needs to be generated (Swaminathan, 1980).

Although there is a pressing need for assuring food for today under no circumstances can we afford to forget to conserve and direct resources for the benefit of the generations yet to be born. Sustained agricultural advances will not be possible without ecological security designed to protect and to conserve fauna and flora, soil and water and environment. In 1980, the World Conservation Strategy was launched under the overall leadership of the International Union of Conservation of Nature and Natural Resources which laid great emphasis on the preservation of genetic variability in living organisms.

The conservation of plants cannot be postponed any longer. The collections can be made in seed, cutting, bulb or any other storage organ or in some cases it may necessitate the removal of the entire individual from the natural habitat—off site. In other cases it may be essential to preserve the plant in its natural habitat—on site.

It is but recently in 1978, under the auspices of the National Bureau of Plant Genetic Resources that a project for setting *in situ Citrus* gene sanctuary was sponsored and financed by the Indian Council of Agricultural Research. The renowned national herbaria of the Botanical Survey of India at Shillong, Howrah and Dehra Dun and the Forest Research Institute, Dehra Dun were intensively studied to review the past work pertaining to *Citrus*. These studies revealed the presence of more than 23 *Citrus* species in India. This study included the information regarding the number of *Citrus* species prevalent particularly in India, their time and place of prevalence, including elevation, climatic features, topography and other characteristics of their natural habitat. Information like local name(s) of the variety, uses and the collector's name and address were also collected. Systematic detailed exploratory surveys were made particularly of the centres of diversity to study the extent of each citrus species. A major emphasis was made on the wild and rare species. The study of mandarin orange particularly was unavoidable.

In view of the fact that the north-eastern Himalayan region is abundantly endowed particularly with the indigenous *Citrus* species, dedicated efforts were made to study this region with vast and varied diverse problems. It was observed that *C. macroptera*, *C. latipes*, *C. ichangensis*, *C. assamensis* and *C. indica* were found in the semi-wild and wild states. These are found in different pockets at varied elevations. All except *Citrus indica* are prevalent and hence do not need any particular attention for their protection. The recent ecological surveys conducted by me in the north-eastern Himalayan region have made it possible to locate an area in Tura Range in Garo Hills where *C. indica*, an endangered species, occurs. This species is the most primitive and perhaps the progenitor of cultivated citrus materials. The importance of wild citrus fruits like *C. indica* can hardly be over-emphasised. These materials, by and large, are adaptable to diverse climatic conditions and could be useful in providing wide genetic stocks for developing varieties resistant to diseases and pests. The investigation revealed that *C. indica* trees in the dense forest had flowers and fruits and were relatively more prolific than those found at the periphery. Closely resembling climbers of this species were also encountered casually in the dense forests. This wild primitive type of citrus species can maintain its character and flourish well in the dense forest. The frequency distribution of this species was recorded in its natural habitat.

In order to allow the populations to follow the same line of natural evolution, on site preservation in the natural environment is essential. It is not possible to restore the whole spectrum of ecological conditions elsewhere if these are removed. The incompleteness of environmental conditions coupled with genotypes that are submitted to continuous alterations by recombination of genetic material, selection will follow different lines from those in nature under self created conditions. These are some of the tangible evidences in favour of *in situ* preservation of variability of *C. indica* in Meghalaya. Detailed studies were carried out to identify an ideal site for preservation of *C. indica* in the natural habitat on a model similar to that of biosphere reserves elsewhere. The Tura Range in the Garo Hills in Mehgalaya has been considered the most ideally suitable for the establishment of the first gene sanctuary in India for *Citrus* in Garo Hills as biosphere reserve. The site was approved by the experts committee.

The site forms a part of the Tura Range which is one of the largest ranges in the Garo Hills in Meghalaya. The area of the site is about 10265.96 ha which has been delineated on maps indicating boundaries. The Garo Hills are characterised by a dense cluster of hills of low elevation with a central ridge traversing from the north-west to south-west known as Tura Range. Many of these hills are below 600 metres in altitude except in Arbella Range where the elevation reaches 960 metres and in the central range which maintains a fairly even altitude of 1200 metres. The highest peak is Nokrek which is 1412 metres high.

The present site of the gene sanctuary as biosphere reserve has the following characteristic features :

1. It is a unique area with unique communities of exceptional interest. It has the centre of distribution of rare or endangered species, *C. indica* in the forest with natural protective barriers.
2. It has a natural genetic diversity for the rare species, *C. indica* which is a primitive one and has the priority.
3. It would be an effective conservation unit.
4. This site is an undisturbed area.
5. It will have long-term legal protection.
6. It will provide opportunity for ecological research education and training.

2

Geography and Climate of India

THE WORD India which traces its origin to ancient Greek literature means land of "Indoi" (Latin *Indus*). Its Persian equivalent refers to Hindu, cognate, with "Sindhu" meaning "the land of great rivers". India is situated midway between the south-eastern countries and middle-east countries of Asia. It lies on the northern fringe of the Indian Ocean, connecting the three continents Asia, Africa and Australia. On the east, west and south,India is surrounded by the Indian Ocean whereas the lofty Himalayas stand on the north. The Indian sub-continent is bounded by Burma in the east, Tibet and China in the north-east and Pakistan in the north-west. The Tropic of Cancer passes through the middle part.

India is the seventh largest country and second most populous country in the world. It covers an area of 32,76,141 sq. km. It measures about 3,219 km from north to south and about 2977 km from east to west. It stretches from latitudes 8°4′ and 37°6′ north and longitudes between 68°7′ and 97°25′ east. Its shape closely resembles a triangle with its apex to the south and base towards the north having the largest continental area. The tapering shape of the country piercing through the Indian Ocean in the form of a wedge divides the oceans into two seas, the Bay of Bengal in the east and the Arabian Sea in the west.

The vast Indian sub-continent has a considerable large range of physical environment. It represents a monumental assemblage of undulating and snow-capped hills; high plateaux and plains have played a dramatic role not only in influencing her history but also her climate, land use communication and distribution of population etc. With regard to physical features, for convenience the country may be divided into four natural regions.

1. The high mountain barrier formed by the Himalayas in the north.
2. The Indo-Gangetic plain.
3. The Deccan Plateau of peninsular India.
4. The coastal strips.

These four natural regions are remarkably distinct in respect of their surface configuration; a brief description follows.

THE HIMALAYAS

The Himalayas and their offshoots cover the entire northern border of the country from the eastern border of West Pakistan to the frontiers of Burma. These Himalayas are the loftiest, uneven mountain ranges in the world. These have enormous height. They consist of three parallel ranges. The greater Himalayan range along the Tibetan border, the Siwalik ranges along the southern margin and the lesser Himalayan ranges in between these two ranges. The great-Himalayan range is the highest range in the world with an average height of 6100 metres above sea level. Mount Everest situated at 8848 metres above sea level is the highest peak in the world. The Siwalik range stretches along the Himalayas west of Bengal Duars. These are generally between 660 metres and 1220 metres high with considerable recent warping. As these hills were raised after the formation of Himalayas, they obstructed the courses of rivers and formed temporary lakes in which the debris of these rivers was deposited. These lakes were drained away wherever the rivers had cut their courses leaving behind plains now called duns e.g. Dehra Dun. The lesser Himalayan ranges are deeply cut by the rivers resulting in the highly rugged mountains. The elevation ranges are generally less than 3050 metres. Apart from the longitudinal sub-divisions the regional characteristics also associated with the Himalayas. The eastern Himalayas are placed relatively near the Tropic of Cancer and near an area of west tropical climate whereas the western Himalayas stretch themselves by about 7° latitude more towards north are located in relatively cool and dry climate. This causes the difference in vegetation.

The Eastern Himalayas

Forests of pine (Pinus khasya) and broad leaved oak are common in the sub-tropical wet climate above a height of 915 metres. The temperate zone starts above 1830 metres high. The monsoon temperate zone persists between 1830 and 2740 metres high. The forests of mixed vegetation e.g. oak, laurel and chestnut particularly of broad leaves are commonly encountered in this zone. The other common trees of this forest are bamboo, ferns and coral. The branches of trees are oftenly covered with the mosses and epiphytes. The damp valleys up to a height of about 915 metres constitute the tropical wet evergreen and semi-evergreen dense and mixed type of forests. The common trees of these forests are "Sal", "Shisham", "Toon" and *Acacia catecheu.* The conifers are found at above 2740 metres height. The prevalent trees found are silver fir, spruce and the common prevalent shrubs are juniper and rhododendron. The alpine vegetation is prevalent between 3655 and 4875 metres in height. The alpines scrub changes into alpine meadows at elevations above 4265 metres. The excellent summer pasture are formed from alpine scrub and meadows during the short summer when the winter snow melts. Stony wastes with lichens and a few herbs are prevalent above the alpine meadows till the permanent snow-line is reached.

The Western Himalayas

The "Sal", "Simal" (Bombax malabaricum), Acacias and Sisoo are useful common trees of this region. Mixed tropical deciduous forests are prevalent in the Garhwal Himalayan foot-hills. The sub-tropical belts prevails between 915 and 1524 metres height. The mixed forest includes a conifer, *Pinus longifolia* (chir) which yields resin and timber. The "Bhabar" grass prevalent at the foothills is used in the paper mills. The conifers "deodar", silver fir, blue pine, walnuts, poplars, maples and oaks are found at varied altitudes in these forests. The temperate fruits, pears, walnuts, apples and cherries are prevalent. The vegetation at higher altitudes is similar to that of the Eastern Himalayas. However, the alpine zone is not found beyond 4570 metres height.

The Himalayas has been divided into four sections on the basis of the rivers (1) Punjab Himalayas—comprises the rivers Indus, Jhelum, Chenab, Ravi, Beas and Sutlej, (2) Kumaon Himalayas—Jamuna, Ganga and Kali, (3) Nepal Himalayas—Karnali, Gandak, Kosi and Tista, and (4) Assam Himalayas—Raidak, Manas and Brahmaputra.

THE INDO-GANGETIC PLAIN

These extensive plains lie in an east-west direction between the Himalayas in the north and the plateau of Peninsular India in the south. They support over half of the total population (approximately 683.8 million). These plains are more than 2400 km in length, with average width of 240 to 320 km. The courses of the rivers are approximately north-south. These plains consist of fertile alluvial soils deposited by three main rivers Ganga, Brahmaputra and Indus. The right banks of these rivers are generally higher than the left banks which are relatively easier sources for irrigation. These plains for convenience are divided into five major parts:

1. The Indus delta, 2. Thar Desert, 3. Ganges Valley, 4. Ganga-Brahmaputra delta, 5. Brahmaputra Valley. Much of the portion of the former two parts falls in West Pakistan. The majority of the portion of the Indus delta which lies within India carries desolate areas, wind blown but still swampy. The Rajasthan (Thar) Desert is primarily covered by a veneer of loose soil and rock materials of varying thickness. The rocks project above the veneer as rocky. The Ganges Valley is the heartland of India. This has very fertile alluvial soil which is very productive and hence is the richest agriculture gift to India. The lower zone along the northern margin of this valley is called Tarai. This area receives the underground seepage from alluvial fans. This potential fertile area is brought into production. The tributaries from the Deccan peninsular region has not much influence in the formation of these plains. Broadly, the Brahmaputra Valley resembles Ganga Valley in surface and structure. This valley has low marshy strip known as "duars". This valley faces relatively more severe problems of floods than the Ganges Valley.

The Ganges-Brahmaputra delta comprises the meeting ground of these two rivers. The surface is uniform, level but low lying. This delta is about 6400 km in length and above 4800 km in width. The other great famous river of this region is the Tista. At the west part of the delta lies the river Hooghly.

THE DECCAN PLATEAU OF PENINSULAR INDIA

The peninsular India lying below the Tropic of Cancer is called the Deccan plateau. The mountain ranges lie on all the three sides of this plateau. This part of India is geologically stable and older than the Himalayas. It is formed primarily of the ancient igneous rocks. It consists of crystalline metamorphosed ancient sedimentary rocks. Some changes in the landscape were brought by the earth movements. The plateau block was uplifted and tilted to the eastward and lines of hills termed "Ghats" were formed at the western edges. These "Ghats" are high and stretch with only a single break, the Phalghat gap. These hills appear as a line of low hills with elevation ranges from about 660 to 990 metres. The high land character is witnessed in the south where the western meet the eastern. In this area, there are at least five high land masses, these are Nilgiri, Vallangiri, Annamalai, Palni and Cardamom hills. The Eastern Ghats are in general lower than the western and are broken up. The Western Ghats are relatively more rugged and are of a lesser wider extent than the Eastern Ghats. The Eastern Ghats have iron ore deposits. The main rivers of this plateau are Mahanadi, Godavari, Narbada, Krishna, Kaveri and Tapti. All the rivers of the Deccan with the exception of the Tapti flow eastwards towards the Bay of Bengal.

THE COASTAL STRIPS

The deposits of numerous rivers and streams draining the plateau have developed a more or less continuous coastal plain at the eastern and western margins of the peninsula. It is a very fertile coast as it is made of soil brought down by rivers. The northern part of the coast down as far as Goa is called the Konkan, and the southern one is known as Malabar. The elevations of these Ghats vary from 815 to 1220 metres.

The eastern coast

The Eastern Coast strip stretches from the delta of the Ganga to Cape Comorin. It is wider in Tamil Nadu than in the western strip. The whole coast is called the Coromandel Coast. The eastern coast is fringed at a few places with dunes. The coast is very fertile and irrigated and consists of numerous deltas formed by the Mahanadi, Godavari, Krishna and Kaveri. These rivers have spread over almost the whole of this plain and, at places large deltas have been formed. The height of this coast does not exceed 610 metres.

The western coast

The Western Coast strip is generally narrow and extends from the Tapti to Cape Comorin. It is a very fertile coast as it is made of soil brought down by the rivers. The northern part of the coast down as far as Goa is called Konkan, and the southern one is known as Malabar. The elevations of these Ghats vary from 815 to 1220 metres.

India has several islands in the Bay of Bengal and Arabian Sea and two fresh water lakes—Wular in Kashmir and Kolair in the south. There are also three well known salt lakes, Chilka and Palicat on the East Coast and Sambhar in Rajasthan. Andaman and Nicobar in the Bay of Bengal and Laccadive, Minicoy and Amindive in the Arabian Sea are the most important islands of the Indian territories. The Sea islands have a tropical climate and thus they have mild winters and summers also.

CLIMATE

Blanford in the later part of 19th century was the first to attempt the systematic study of the Indian climate. Later, Roppen divided the country broadly into three climatic regions—arid, semi-arid and humid. Subsequently more viable classifications were developed by the induction of more attributes like duration of wet and dry months.

India has a variety of distinct climates in different parts of the country. There exist gradual yet complete transitions between the opposite extremes ranging from dry to excessive wet and from intense heat to extremely cold. The Tropic of Cancer passes through the middle part of the country and hence the major part of the country (southern part) lies within the tropics and the rest of it is sub-tropical. The latitude, altitude, soil structure, the distance to the sea and direction of winds and mountain ranges are the six important factors that play a significant role in determining the climate of the country. The places near the equator are hotter than those places further in the north. So are the plains in comparison with the mountains. The places near the sea have a more equable climate as the sea has the capacity to equalize the temperature. Whereas the desert of Rajasthan having a sandy soil has relatively extreme temperatures during the day and night. Similarly, the Himalayas play a great role in determining the climate by preventing the cold winds of the north from reaching India and the moist winds of the Indian Ocean from passing out of India. The other similar evidence is the Western Ghats, running from north to south which get more rains than the Satpura or Aravalli ranges which run in another direction. The winds play a major role in affecting the climate of the country. The surface wind that blows in a particular direction but reverses in the other is known as "monsoon" of which there are two types.

RAINFALL

South-west monsoon

India is essentially a monsoon land. The monsoon has a great significance in governing Indian life, as India primarily depends upon agriculture. The word monsoon traces its origin from the Arabic *mausim* which referred to shift of winds with the change of seasons in the Indian Ocean. The concept of the word *mausim* attached great importance to the Arabs who used the Indian Ocean as their trade route to India and adjoining countries during the summer months. The effect of the seasonal shifts were subsequently realised by the Portuguese, Dutch and British traders.

The south-west monsoon comes from the sea and hence brings plentiful rains which make the climate moist. The thermal changes are the main causes associated with the monsoons. In the northern hemisphere, the sun's rays falling perpendicularly near the Tropic of Cancer in the summer months heats the air which rises due to expansion and causes decrease in the pressure. Recently it is believed that jet streams and the topographical barrier of northern India are also two other major attributes to influence the monsoon. The high altitude mountains of Central Asia isolate partially South Asia from the rest of the continent. The high elevation associated with gigantic size alter the flow of jet stream which normally occurs above the latitudes of 20° to 40° N.

The summer season in general can be divided into two parts. (1) Pre-monsoon rainy season, and (2) the hot and dry season, which prevails from March to May in India except in north-western India, where it prevails from April to June. It is characterised by aridity and excessive heat. The temperature increases as the summer advances in May to early June. The prevailing daily maximum temperature is over 38°C in the mid-Ganges area and it goes over 49°C in the desert of Rajasthan which is the hottest part of Asia during this period. During the same time an average high temperature of over 35°C prevails in the Deccan plateau.

The low pressure exists in the Indo-Gangetic plain coupled with relatively high pressure in the Indian Ocean. This causes the wind to blow from the Indian Ocean to the heated plains where there is low pressure. These winds normally blow somewhere from June to September in the south-west direction and hence called South-West Monsoon. This summer monsoon has low surface air currents: (1) Bay of Bengal current, blowing over the Bay of Bengal and (2) Arabian Sea current, blowing over the Arabian Sea and the Indian Peninsula.

A part of the current of Bay of Bengal passes over the Ganga delta and strikes Khasi, Jaintia and Garo Hills. The southern slopes of the Khasi Hills (Shillong Plateau) receives tremendous amount of rainfall as a result of sudden rise of the monsoon to a height of above 1600 metres. The monsoon winds sweep up the length of the Bay of Bengal gethering moisture as they proceed. The Cherrapunji area is the rainiest place on the surface of earth. The other part of the current of the Bay of Bengal branch strikes the lofty Himalayas and unable to pass beyond, passes through the Ganga Valley and becomes the south-west monsoon. As it passes from east to the west, it gradually loses the moisture resulting in a decrease in rainfall. Similarly there is more rainfall at the foothills of the Himalayas i.e. in the Tarai situated on the northern side of the Indo-Gangetic plain than in the further south.

North-east monsoon

The current of this monsoon blows from November to February. The sun's rays falling obliquely in the north and north-west part of the country brings down the temperature. Obviously, as a result, there is higher atmospheric pressure than in southern India and on the Indian Ocean. This causes the movement of the wind from the north-west along the valleys of the Ganga and the Indus and this becomes the north-east or winter monsoon. It is drier since it comes from the land. The part that passes over the Bay of Bengal picks up moisture and brings rain to the eastern coast. Places like Madras receiving both the monsoons have the advantage of two rainy seasons.

Distribution of rainfall

In general, broadly there are two regions : one of heavy, the other of scarce rainfall.

Region of heavy rainfall

1. The Assam hills and the Brahmaputra valley.
2. The Ganga Valley from the delta to Amritsar in Punjab.
3. The Tarai area at the foothills.
4. The West Coast from Tapti to Cape Comorin.
5. The East Coast from Ganga delta to Cape Comorin.

The regions of scarce rainfall

1. Western Rajasthan desert.
2. The Deccan.

It is observed that there is general decrease in rainfall from east to west; Calcutta receives 158.0 cm whereas Delhi receives 68.0 cm. Calcutta starts getting the effect of monsoon by the middle of May whereas in Delhi the rains begin in mid June. It is significant to note that one-third of the year which is the south-west monsoon period receives over 80% of the year's precipitation in mid-Ganga Valley up to the Punjab. During the same period Ganga-Brahmaputra delta receives about 66%.

THE SOIL RESOURCES OF INDIA

A thin surface layer of collection of natural bodies arising from the disintegration of rocks into finer particles over a period of time that can support life is called soil. The Indian soils have generally developed under a hot and humid climate over the bed rocks predominantly by the process of laterization, though podzolisation also occurs. Under the soil, there is accumulation of leached materials on the gravels of parent rock, collectively known as sub-soil.

A large portion of humus is destroyed due to the active decay of dead vegetation resulting from high prevalent temperatures during most of the year in India. It is obvious that a vast majority of the soils of India are deficient in humus and need continuous application of fertilizers and manures that are particularly rich in nitrogen. The soils formed from parent rocks consisting of limestone are likely to be alkaline. The soils are relatively less productive and shallow on the steep slopes wherein the soil is regularly washed away. The appropriate contour bunding reduces the erosion of soil. The strong winds such as the south-west monsoon blowing over arid areas like Rajasthan, carry away fine dust, thus leaving behind infertile soil comprising large sand particles. The soils of coarse particles are termed as sandy soils e.g. desert of Rajasthan, whereas the soils of fine particles are classed as clays. The soils consisting of intermediate size between clay and sand are called silt. Soils formed by the admixture of the three types of particles i.e. clay, sand and silt are called loam. The sandy soils are often called light soils. They are well drained with poor holding capacity of nutrients and water and hence need continuous application of nutrients. The clay forms generally heavy soils. These are sticky and hence relatively difficult to plough. They are generally rich in nutrients but are liable to water-logging. Silt and loam forming intermediate textures are naturally better than clayey and sandy soils. These have the moisture retention capacity and also do not suffer much from water-logging. These are more productive. The soils of the plains of northern India are commonly loamy and silty.

The Soil Survey of India was established recently in 1956. There is no complete agreement in the procedure of classification, with the result it has not been possible to reconcile many soil maps. The detailed accurate soil maps are still awaited. Broadly, for convenience, the soils of India are divided into five groups : alluvial, laterite, regur, mountain and desert.

Alluvial soils

These soils are formed from rivers and are an admixture of finer gravel, sand, silt and clay. These are generally excellent loams which are rich in nutrient with good water holding capacity. These are easy to plough and are not exposed to water-logging conditions. These soils are in general, very productive and hence are the support of large number of Indians. These soils stretch over large area of the plains of Northern India which includes Uttar Pradesh, Bihar, Punjab, West Bengal and part of Assam and Orissa. In the Indian Peninsula, these are confined to the deltas on the east coast, the lower valleys of the Narbada and the Tapti, Chattisgarh Plains and Northern Gujarat. Mostly they have low gradient and unlike other soils have not suffered much soil erosion. These soils are generally rich in lime and potash but poor in humus, nitrogen and phosphoric acid. Due to the inconsistency of river depositions, banks of gravels, sand bars or pure clay are found which are of poor productivity. The lowlands soils of Brahmaputra and Ganges seem to be yellowish whereas the peninsular fringes are reddish.

A considerable area of the Brahmaputra Valley, West Bengal and northern Bihar is leached due to the high amount of precipitation. The soils have become deficient in calcium due to the depletion of lime. The soils of the Assam Hills are acidic and so are the soils away from the rivers. Though due to the deposition of fresh alluvium near the Brahmaputra, the soil may not be deficient in calcium. In the north-western drier parts which receive low rainfall, the soils have a tendency to accumulate alkaline salts composed of calcium, sodium, potassium and magnesium in their upper layer. These soils are commonly called 'kallar' or 'reh'. These alkaline soils are poor in productivity. The alkaline soils are commonly found in semi-arid conditions where there is excessive evaporation, the salts are brought to the surface layer and left behind. A similar situation is met with in many areas fed by canal waters, where due to the high water table, there is accumulation of alkaline salts in the top layers of soil resulting from the evaporation. Depending upon the extent of alkalinity some of the crops like rice, sugarcane, alfalfa, jowar and barley can be successfully grown on such soils.

Lateritic soils

These soils are formed by the withering of laterite rocks. These are distinctively acidic soils and are found in tropical regions receiving heavy rainfall. The high amount of precipitation encourages leaching silica and lime leaving behind the oxides of iron and aluminium compounds. The iron oxide formed by the combination of oxygen with iron is reddish. Laterite is called bauxite in places of abundance of aluminium oxides. These soils have poor water holding capacity and are also poor in productivity. These soils are definitely less productive than either regur or neighbouring alluvial soils and are sometimes barren. In low lying areas where there are chances of additions of clay or loam, the soils are relatively fertile but need regular applications of nutrients for crop production. These red lateritic soils are primarily derived from ancient sand stones and are sandy and poor in humus. These soils cover a vast area from Delhi to Cape Comorin and north-eastward to Chota Nagpur and Shillong. They are found in Andhra Pradesh, Orissa, eastern and southern portions of the Western Ghats and the extreme tip of the peninsula, in scattered patches of Chota Nagpur, West Bengal Assam and all of the Shillong plateau. The effect of strongly acidic soil conditions is reduced in the areas of high precipitation where the silica portion of clay materials of soil are leached away leaving behind the iron, aluminium and manganese components. The alternations between dry and wet seasons play a dominating role in the formation of the hard crust of the laterites.

Regur or black soils

These soils are black in colour and are the product of decomposition of lava. These are developed under the semi-arid conditions in areas covered with basalt. The basalt lava, high accumulation of calcium in the upper layers, original savanna grass land vegetation and deep oxidation are some of the potent reasons ascribed for its black colour. The regur soils with black colour are conspicuous which are in sharp contrast to alluvial, desert and lateritic soils. These soils have the extremely fascinating property of self-aeration. The soil develops cracks during the dry season when it loses moisture. The usefulness of these cracks lies in aeration and also they enrich the soil by pushing the fallen vegetational material into lower levels wherein these are incorporated within the soil. Thus, these soils have developed a natural system to maintain their productiveness. These soils are known for their stiffness and stickiness when wet. This characteristic is due to the high proportion of clay, and this results in considerable difficulty in ploughing. These soils are generally fairly rich in plant nutrients and are valued for their ability to retain moisture for longer periods in spite of high surface evaporation. These soils are generally rich in lime and potash but are deficient in nitrogen and phosphoric acid. The regur or black cotton soils are prevalent in Maharashtra, west of Madhya Pradesh, Gujarat, western part of Andhra Pradesh, south of Uttar Pradesh, south-east of Bihar and north-west of Orissa.

Mountain and hill soil

The topographic, geologic vegetational climate conditions are exceedingly diversified and so complex that it is an uphill task to discuss the mountains and hill soils collectively. However, a few generalizations to this complex discussion could be described: firstly, the large area of homogeneous soil does not exist in the mountains; secondly, the mountain soils are commonly immature and shallow because of the slopes; and thirdly, a large area of mount soils are gravelly, and acidic in nature. We do have some local alluvial or lacustrine deposit in basins.

Desert soils

These are generally grayish or tan coloured soils and are characterized by low organic matter coupled with a lower rate of decomposition of humus. These are commonly found in Rajasthan, part of Punjab and Haryana states which fringe Rajasthan. These are sandy soils. These are also prevalent along the river courses and along the coastal strips. These soils are poor in nitrogen but fairly rich in other plant nutrients. They produce rich crops under irrigation conditions. The fertility of such soils is reduced considerably by the strong dust storms which carry fine dust with them.

NATURAL VEGETATION

As the present investigations pertain to the setting of the *Citrus* Gene Sanctuary in its natural habitat in the forests on a model similar to the biosphere reserves, it is, therefore, essential to give a brief description of the nature of prevalent forests in India.

Early man was considered as the integral part of the components of the biosphere and his impact on nature was indistinguishable. But in the recent past the social and cultural evolution of man has been marked by a growing conflict between him and nature. This has resulted in modifying remarkably the vegetation of India over large areas. The geographical area and its classification in respect of major states of India is indicated in Table 1. There remains little undisturbed area. A little over one-fifth of the country is forested. The area under forests in different States/Union Territories of India is listed in Table 2. Most of the large forest blocks occur in the Brahmaputra Valley, Burma border ranges, Western Ghats and Chota Nagpur. The resources of these forests are being carefully utilized.

The destruction of shrubs and grasses as a consequence of overgrazing in drier areas in the west and in the Deccan plateau has posed a serious problem by exposing these areas to soil erosion. Shift cultivation is a common practice in the north-eastern Himalayan region resulting in the mountain ecosystem being denuded of vegetation. The tribals there find it absolutely necessary to clear the forests by uprooting and burning them for raising crops for their very existence.

TABLE 1 : **Geographical area and its classification in respect of major states during 1969-70**

(Thousand Hectares)

State	Geographical Area	Reporting Area for land utilisation statistics	Classification of Reporting Area: Percentage to geographical area	Forest	Area not available for cultivation	Other uncultivated land excluding fallow lands	Fallow lands	Net Area sown
1	2	3	4	5	6	7	8	9
Madhya Pradesh	44,284	44,237	99.9	14,437	4,426	5,558	1,533	18,283
Rajasthan	34,222	34,107	99.7	1,342	5,883	8,224	5,563	13,095
Maharashtra	30,776	30,768	100.0	5,408	2,456	2,255	2,187	18,462
Uttar Pradesh	29,441	29,860*	100.0	4,969	3,460	2,724	1,350	17,357
Andhra Pradesh	27,682	27,440	99.1	6,369	4,135	2,698	2,728	11,510
Jammu & Kashmir	22,224	4,523	20.4	2,778	543	386	117	699
Gujarat	19,598	18,562	94.7	1,634	4,827	1,591	1,082	9,428
Karnataka	19,177	18,938	98.8	2,868	1,775	2,572	1,526	10,197
Bihar	17,388	17,330	99.7	3,017	2,589	895	2,434	8,395
Orissa	15,578	15,540	99.8	4,960	1,802	1,961	723	6,094
Tamil Nadu	13,007	13,004	100.0	2,016	2,327	982	1,610	6,069

West Bengal	8,875	8,852*	100.0	1,101	1,272	607	330	554
Assam	7,852	7,795	99.3	2,080	2,555	644	290	2,226
Himachal Pradesh	5,567	5,082	91.3	2,788	289	1,390	62	544
Punjab	5,036	5,030	99.9	119	635	99	150	4,027
Haryana	4,422	4,402	99.5	97	488	109	161	3,547
Kerala	3,886	3,859	99.9	1,056	842	249	46	2,166
Other States/Union Territories.	19,835	16,713	84.3	9,005	6,345	284	25	1,054
All India	328,758	306,042	93.1	66,044	46,158	33,228	21,917	138,695

*Under verification.

Source : (Anonymous, 1977). Indian Agricultural Statistics, 1967-68 to 1969-70, Vol, II. Directorate of Economics & Statistics, Ministry of Agriculture & Irrigation, Govt. of India.

TABLE 2 : **Area under forests in States/Union Territories during 1969-70 according to the agricultural statistics returns.**

	Area under forests (Thousand hectares)		Percentage to total
Madhya Pradesh	14,437	(32.6)	21.9
Andhra Pradesh	6,369	(23.2)	9.6
Maharashtra	5,409	(17.7)	8.2
Uttar Pradesh	4,969	(16.7)	7.5
Orissa	4,960	(31.9)	7.5
Bihar	3,017	(17.4)	4.6
Karnataka	2,868	(15.1)	4.3
Himachal Pradesh	2,788	(54.9)	4.2
Jammu & Kashmir	2,778	(61.4)	4.2
Assam	2,080	(26.7)	3.1
Tamil Nadu	2,016	(15.5)	3.1
Gujarat	1,634	(8.8)	2.5
Rajasthan	1,342	(3.9)	2.0
West Bengal	1,101	(12.4)	1.7
Kerala	1,056	(27.4)	1.6
Punjab	119	(2.4)	0.2
Haryana	97	(2.2)	0.2
Other States/ Union Territories	9,005	(53.9)	13.6
All India	66,044	(21.6)	100.0

Note : Figures in brackets indicate the percentage of forest area to reporting area of the State.

Source : (Anonymous, 1977). Indian Agricultural Statistics 1967-68 to 1969-70, Vol. II, Directorate of Economics & Statistics, Ministry of Agriculture & Irrigation, Govt. of India.

The climate of India is essentially tropical. As a result tropical vegetation up to a height of 915 metres above sea level is found throughout the country. Altitude, rainfall, temperature, soil, slope of land, interference of man, these are some of the potent factors which influence to a considerable extent the natural vegetation. The uneven rainfall, varied edaphic and biotic conditions have given rise to as many as 30,000 species of plants in the country ranging from thorny bushes to evergreen forests. The important families to which the forest vegetation belongs are : Orchidaceae, Euphorbiaceae, Acanthaceae, Grammineae, Leguminosae, Cyperaceae, Labiatae,

Urticaceae and Compositeae. There are several factors like climate, soil, species single or in combination which determine the character of a forest. These have been divided into five groups and the description of these follows.

Mangrove swamp forest

The mangrove forests are found along the coastal fringes of the Ganges-Brahmaputra delta. These forests also appear in thickets on the western and eastern coasts along the delta fronts of the Ganges, the Mahanadi, the Krishna and the Godavari and along the coast of the Andaman Islands. These forests flourish well on the muddy and swampy coasts adjoining the deltas. These are also found in the Ganga Delta fresh water swamps where semi-evergreen forests exist comprising bamboos, palm, ferns, canes and tall grasses. These forests contain Sundri trees (Heritiera fomes) and hence these are commonly known by the name of 'Sundarbans'. The mangrove forests are associated with tidal floods and grow in profusion in the salt water. The trunks of the trees are commonly found supported by stilt like roots to fix the trees in the soft mud banks. The tangle of climbers associated with dense profuse growth makes these impenetrable. The Sundri trees of the Sunderbans are an important source for providing wood for small boat construction.

Tropical rain forest of Selva

These forests are found in the Ganga-Brahmaputra delta in the Assam Himalayan region and in the Western Ghats and Andaman Islands. The tropical forests are well developed particularly in the Brahmaputra Valley and on the slopes surrounding the delta. The profuse growth of the trees makes them grow high competing with each other for sunlight and these arrange themselves like the trees in the Selvas into three or four storeys. The tropical wet conditions favour the growth of an enormous number of plant species. The trees of these evergreen forests are generally of hardwood, durable and fine grained. 'Champa', 'toss' and 'gurjan' in the Himalayas and rosewood, 'poan', 'aini' and 'telsur' in the Western Ghats are some important commercial timber trees of these forests. A number of epiphytes climbing over the trees in this forest are also encountered. A variety of fascinating orchids flourishing on these trees are a special feature of this region. It is common to find a dense undergrowth of palms, bamboos, canes, ferns and a tangle of climbers near the margins of the forests.

What is remarkable to note, however, is that Cherrapunji and its enormous adjoining areas receiving a world record rainfall are deprived of trees. The probable causes for this phenomenon may be attributed either to torrential rainfall which washes away soil and thus prevents trees from establishing roots or the reason may be biotic. Whenever there is relatively low rainfall the evergreen forests degenerate into semi-evergreen forests.

Tropical deciduous or open monsoon forest

The open monsoon forest covers an enormous area of Central India, the swampy Tarai zone on the northern margin of the Ganges Valley, Chota Nagpur and a part of Siwalik hills. A considerable area of these forests have been reduced due to the obvious reasons of civilization, cultivation and grazing. The precipitation in these forests varies from 152 to 203 cm. The tropical monsoon forests contain a limited number of species with an undergrowth of bamboo and cane climbers and epiphytes. The luxuriant vegetation degenerates in areas of scanty rainfall below 152 cm and during dry spells. In these areas, tropical dry deciduous vegetation is found which has relatively short shrubby and grassy undergrowth. The patches of natural and artificial savana grasslands are the characteristic of these areas. The most common trees of these forests are *Tectona grandis* (teak) in the central and western, *Santalum album* (sandal) in the southern and *Shorea robusta* (sal) in the north-east uplands of the Indian Plateau and in the tropical Himalayas. The sal and teak are very valued timber trees of the country. The other common trees of these forests are *Dalbergia sisoo* (shisham) and *Terminalia chebula* (hurra), *Bassia latifolia* (mahua) and *Acacia catechu* (khair).

Thorn forest and scrub

There are two zones characterizing these forests. The thorn forest stretches from the Tapti Valley to the Coromandel coast and in the interior of the south. The other scrub belt is spread over Kutch and the neighbouring parts of Kathiawar and on the eastern fringes of Rajasthan including Aravalli hills, western Haryana and western Punjab.

The vegetation in these belts is thick and dark green during the monsoon whereas in the dry season it withers and becomes dusty. The trees in general are stunted and bushes are found in abundance. These forests have some xerophytic undergrowth. The most common thorny trees of here are Babul (*Acacia arabica*), *euphorbia*, *mimosae*. The other prevalent plants of these belts are Khair (*Acacia catechu*), beri (*Ziyzhus*), canes (*Sacharum spontaneum*) and oak. There are some palatable grasses which grow during the rainy season. These are used as cattle feed.

Mountain vegetation

Mountain vegetation is of several types due to the varied climate from tropical heat to heights of arctic cold. The elevation is one of the most important factors influencing the vegetation. Varied type of vegetation is encountered : evergreen vegetation at the foothills; and at heights from 1524 to 3658 metres temperate forests and at still higher elevations, alpine vegetation is found. The Eastern Himalayas are stretched near the tropic of cancer and near the wet tropical climate whereas the Western Himalayas are placed by about 7° latitude more towards the north in the relatively dry and cool climate of the continent. This results in a change of the vegetation characteristics of the two parts of the Himalayas.

The tropical rain forest is prevalent up to a height of about 1300 metres above sea level. A zone of partially deciduous forest is found above this elevation. Sal, oak and chestnut are prevalent. Rainfall and exposure are two other important factors to differentiate vegetation. The sub-tropical forest with broad leaf vegetation changes to sub-tropical pine forest at about the same height westward along the Siwaliks and in the lower slopes of the Himalayas. A similar type of vegetation exists in the Shillong plateau. In west Kashmir which is drier in the lower slopes, the vegetation grades from sub-tropical pine to sub-tropical dry evergreen. The most common trees of this belt are acacia, dwarf treepalm and wild olive.

The temperate forests are situated above the sub-tropical forests. The wet temperate forest stretches from east up to about 80° E. The vegetation in general is dwarf, the common plants of the region are chestnut, oak, laurel and rhododendron. The deodar trees in their pure state exist west of 80° E in the forests. The temperate forest assume xerophytic characters in areas of below 100 cm rainfall.

Desert shrub

This type of desert shrub is prevalent in India in the Thar Desert and behind the Himalayas on the plateau of Tibet. In this zone, the trees are stunted, twisted, and reduced to shrub. These have adapted themselves to conserve moisture. The vegetation in this belt develops xerophytic characters such as waxy leaves, water-storing stems, spines, thorns, reduction of pores and transpirational surfaces. However the vegetation growth is stunted due to the dryness and cold temperature prevalent at the high elevations placed behind the Himalayas on the plateau of Tibet.

3

Origin of Citrus Fruits

THE CITRUS fruits which include oranges, lemons, limes and citrons rank third in acreage among the sub-tropical fruits of the world. These fruits are grown in tropical and sub-tropical regions throughout the world and have been cultivated in South China, Malaya and the sub-Himalayan parts of erstwhile Assam from time immemorial. Subsequently these spread to the tropical and sub-tropical parts of the world. Though the citrus fruits are not included in the list of those which have been in cultivation for 4,000 years by De Candolle, it seems that it was known to the Chinese at least that long. Hu (1934) states that a mention of citrus has been made in the Chinese literature written as early as 2200 B.C. Citrus fruits were exchanged between India and China before the earliest records, through usual trade routes.

Of several kinds of plants for the gardens and orchids of the world, there are none more important than those of the citrus group which have been contributed by China. North Central America is the leading producer with over 33% of the total acreage in the world under citrus fruits. The U.S.A. and the Mediterranean countries lead in their production. Florida alone produces 75% of the world's grape fruit, Italy leads in lemons whereas India and Mexico lead in limes. Asia is the second largest producer in the world for lemons, limes, grape fruit and pummelo (Table 3).

The total number of *Citrus* species in the world ranges between 16 and 157 but only 37 of these are cultivated in India. It has been stated unequivocally that the citrus fruits are native to Southern Asia. Hays (1966) stated that some citrus fruits originated from China and some from India while others

TABLE 3 : **Production = 1000 MT**

	Oranges				Tangerines, mandarins, clementines and satsumas			
	1969-71	1976	1977	1978	1969-71	1976	1977	1978
1	2	3	4	5	6	7	8	9
World	25453	33964	34338	34110	5162	6816	7356	7036
Africa	2688	2908	2754	2949	388	423	400	426
NC America	9242	11901	12515	11636	615	764	748	715
South America	4915	9108	9140	9726	577	690	738	741
Asia	4433	5181	5290	5455	2885	3797	4296	3878
India	900	970F	1004F	1037F	—	—	—	—
Europe	3812	4404	4055	3775	672	1105	1141	1242
Oceania	291	333	359	369	25	37	33	34
Australia	283	322	347	357F	21	31	26	28F
USSR	72	128	226	200F	—	—	—	—

TABLE 3—(*contd.*)

	Lemons & Limes				Grape fruit & Pummelo				Citrus fruits NES			
	1969-71	1976	1977	1978	1969-71	1976	1977	1978	1969-71	1976	1977	1978
	10	11	12	13	14	15	16	17	18	19	20	21
World	3672	4476	5038	4645	3339	4059	4304	4228	588	859	886	895
Africa	217	243	209	218	198	207	212	223	242	301	310	306
NC America	578	1196	1504	1444	2318	2753	2912	2874	65	78	81	84
South America	433	518	615	602	197	249	243	226	116	227	212	215
Asia	933	1198	1257	1256	597	807	888	856	160	247	275	283
India	450	450F	451F	453F	20	20F	20F	20F	28	28F	28F	29F
Europe	1078	1278	1412	1082	12	14	19	20	1	2	2	2
Oceania	33	42	40	42	18	29	29	30	3	5	5	6
Australia	28	36	35	36F	13	21	22	23F	—	—	—	—
USSR	—	—	—	—	—	—	—	—	—	—	—	—

F = FAO estimates
MT = Metric Ton
NES = Not elsewhere specified or indicated
Source : FAO Production Year Book, Vol. 32, 1978

might have originated in the regions in between these two old countries, the Malaya Archipelago, new Guinea and Australia. The north-eastern Himalayan region holds a notably important position in regard to citrus wealth.

The north-east region, south India and the north-west region at the foot of the Himalayas are the three regions which are considered the major centres of diversity in the country. Some indigenous and wild mandarin types are found in south India. Whereas the hill-lemon (galgal) and 'attani' are prevalent at the foothills of the Himalayas in the north-west. The north-east region is endowed with remarkable genetic richness in a number of plant species i.e. rice, maize, sugarcane, jute, banana and *Citrus*. However, for some of the crops like *Citrus*, banana and mango, the natural diversity prevails in the forests.

The north-eastern Himalayan region offers a favourable agro-climatic conditions for the flourishment of different citrus fruits and is considered the natural home of several *Citrus* species. The papedas, pummelos and their hybrids, *C. Indica* and many types of citrons, lemons and mandarins are found in this region. Bhattacharya and Dutta (1951) reported 17 species comprising 52 varieties in this region. They have reported that *C. indica*, *C. assamensis*, *C. latipes*, *C. ichangensis* and *C. macroptera* are indigenous to Assam. *C. microptera* is also claimed to be a native of Thailand. Singh (1957) reported that *C. aurantium*, *C. megaloxycarpa*, *C. reticulata*, var 'Soh-syiem' are also considered to be indigenous to Assam. Singh (1977) reported the species which are admittedly indigenous to India include *Citrus indica*, *C. latipes*, *C. megaloxycarpa*, *C. karna*, *C. Jambhiri*, *C. aurantifolia*, and possibly *C. aurantium*, *C. medica*, *C. sinensis*, *C. madurensis* and *C. limonia*.

The indigenous mandarin is mostly related to the pummelo, the citron and the karna khatta, which are monoembryonic or have very weak polyembryony. Though a remarkable number of varieties are found in monoembryonic fruits, it seems several types are prevalent among the mandarins which are polyembryonic. The cultivars of *Citrus limon* Burn (lemon), *C. medica* (citron), *C. aurantifolia* Swingle (lime), *C. grandis* (pummelo) and *C. reticulata* are grown to a considerable extent in this region. Practically one or the other variety of citrus is growing in every corner of India.

The loose skin orange—Khasi mandarin (*C. reticulata*) is the speciality of the region. Ghosh (1977) states that Assam alone exports oranges worth 50 lakhs of rupees every year. The main mandarin growing belts in this region are Kamrup, Dibrugarh and Lakhimpur districts in Assam; Changki valley of Mokokchung and certain areas of Tuensang districts in Nagaland; Tamanlong, Tipaimukhs and Jaribam areas along with the belts of the Barak river in Manipur; Jampui Hills area in Tripura and Shella area in the southern slope of Khasi Hills and Dawki, Jowai, Shanpung in Jaintia Hills in Meghalaya. The estimates of area and production of citrus fruits in this region are given in Table 4, and the local names of some of the citrus fruits have been indicated in Table 5.

TABLE 4 : **Area and production of citrus fruits of the States/ Union Territories of North Eastern Region**

State/Union Territories	Area '000 Ha	Production '000 Tonnes	Yield Kg/Ha	Year
Meghalaya	6.51	45.00	6914	1974-75
Manipur	6.07	43.00	7084	1975-76
Assam	5.75	50.21	9219	1972-73
Mizoram	3.00	11.50	3833	1973-74
Tripura	2.47	8.57	3469	1973-74
Arunachal Pradesh	—	0.16	—	1973-74

TABLE 5 : **Local name of some of the citrus species of North-Eastern Himalayan Region**

Botanical name	Khasi/Garo/Jaintia name	
C. reticulata	Soh niamtra (Khasi Mandarin)	Khasi name
C. jambhiri	Soh myndong	Khasi name
C. macroptera	Soh kwit	Khasi name
C. assamensis	Soh sying	Khasi name
C. grandis	Soh myngor	Khasi name
C. latipes	Soh kymphor shrieh	Khasi name
C. indica	Memang narang	Garo name
	Sa kymphria	Jaintia name

Of the total area of nearly 11 lakh hectares under the different fruits in India, over 1.0 lakh hectares are occupied by the citrus fruits (Table 6). The mandarin oranges are commonly called 'santra', the lime and lemon are grown extensively, whereas some of the citrus fruits like grape fruits are grown to a limited extent. Area of citrus fruits in different States/Union Territories of India is indicated in Table 6.

TABLE 6 : **Area of citrus fruits in different States/Union Territories of India (1967-68 to 1969 70)**

Area in Hectare

S No	State/Union Territories	Year 1967-68	1968-69	1969-70
1.	Andhra Pradesh	27,704	29,205	30,285
2.	Assam	5,014	5,073	5,070
3.	Bihar	—	—	—
4.	Gujarat	3,100	3,000	5,500
5.	Haryana	2,250	2,155	1,428
6.	Himachal Pradesh	657	185	223
7.	Jammu & Kashmir	—	101	57
8.	Karnataka	10,517	9,433	11,877
9.	Kerala	1,959	1,959	1,959
10.	Madhya Pradesh	5,582	5,768	5,873
11.	Maharashtra	25,374	24,900	25,400
12.	Manipur	—	—	—
13.	Meghalaya	4,013	4,013	4,012
14.	Nagaland	—	—	—
15.	Orissa	1,800	1,800	1,800
16.	Punjab	636	5,709	952
17.	Rajasthan	2,251	3,494	3,150
18.	Tamil Nadu	2,942	2,828	3,990
19.	Tripura	2 365	2,400	2,450
20.	Uttar Pradesh	131	146	76
21.	West Bengal	—	—	—
22.	Andaman & Nicobar Islands	24	35	35
23.	Dadra & Nagar Haveli	—	—	—
24.	Arunachal Pradesh	—	—	—
25.	Delhi	99	—	—
26.	Goa, Daman & Diu	—	—	—
27.	Mizoram	947	945	946
28.	Pondicherry	8	1	—
29.	Lakshya Deep	4	4	4
	Total	97,377	103,154	105,087

Source : (Anonymous, 1977). Indian Agricultural Statistics 1967-68 to 1969-70, Vol II, Directorate of Economics & Statistics, Ministry of Agriculture and Irrigation, Govt. of India.

There is considerable difficulty with the popular definition of oranges and lemons. The 'Khatta' citrus has orange coloured leaf, exterior pulp and yet its fruit is prominently mammillate and its flower is tinged purple. Similarly the 'Mitha nimboo' which is acidless citrus has a lemon-yellow exterior, white pulp and is mammillate with lemon-like leaves, yet its flowers are white. The 'Jambhiri' is another example having mammillate fruit orange or lemon-yellow coloured externally yet it has orange like leaves with slightly tinged purple flowers.

There are two types of sweet oranges : the loose-skinned orange and close skinned Portugal orange. The trees have distinct characteristics. The loose skinned has existed from a remote period in the Eastern Himalayas while the Portugal orange seems to be a relatively recent introduction in India. The former is found under various names all over India. It was known as 'China orange' in the Malay Archipelago in the time of Rumphius. It is prevalent in a semi-wild state in the Eastern Himalayas, and might be inferred as indigenous to India.

It appears that in Sanskrit and also in European languages the names which were first applied to bitter orange were in later times appropriated for sweet oranges. 'Nagaranga' in Sanskrit and 'Narangi' in Hindi are the equivalent words of 'Narang' the Persian word meaning orange. It has not been possible so far to derive satisfactorily from this name that it is indigenous to India. 'Nar' means fragrance in Tamil. The Séville orange or the *Citrus bigaradia* of Risso and the 'narang' of the Arab are probably one of the most ancient varieties of citrus.

The north-eastern Himalayan region which was once known to be a rich reservoir of genetic wealth of citrus has lost considerable number of citrus fruits and the situation is alarming due to socio-economic changes.

There has been a lot of confusion in describing oranges. Bonavia (1890) has divided the sweet oranges into two sections one 'Metha' or Portugal orange—*C. aurantium sinense* of Gallesio, and its varieties and the other 'Suntara orange of India' the *C. aurantium sinense* of Ramphius and its derivatives. The metha or Portugal orange does not have a slender tree and aromatic leaf but has distinct close skin orange fruit.

The Suntara orange of India has a distinct slender tree with aromatic leaves and loose skin orange fruit. These are sour when unripe, the acid changes to sugar as they ripen. None of these is totally sweet. Acidless oranges are not found in India. He further states that the Séville orange is not bitter but is simply sour. It does not have a bitter pulp. He has reported one variety of Séville orange as 'Devikulam' in south India, but possibly nothing could be said about its origin, whether it is indigenous or has become naturalized there.

It does not seem probable that the Séville orange (sweet) is an Indian orange. Firstly the climate of India is not suitable for its growth; secondly, it has occurred under several names, most of them being names of other kinds of citrus. It is called 'gora lemboo' in Calcutta 'sadaphal' in Gonda, 'naranj' in Lucknow, 'karna' or 'kanchikai' in Mangalore and Mysore and 'khatta' at certain places. Alphonse de Candolle in *Origin of Cultivated Plants* mentions, "Oranges are distinguished from Shaddocks by the complete absence of down on the young shoots and leaves, by their smaller fruit, always spherical and by a thinner rind. They differ from lemons and citrons in their pure white flowers, in the fruit which is never elongated and without a nipple on the summit, in the rind being smooth, are nearly so and adhering but lightly to the pulp". Bonavia (1890) indicated that the characters mentioned by De Candolle do not fit in with the reality. Seemingly, it is difficult at times to distinguish an orange, a pummelo, a lemon or citron. It is often considered that the orange should have white flowers and red spherical without mammilla fruits whereas lemon has red flowers, is yellow elongated with mammilla fruits. This does not seem to be the natural reason. This situation could be compared with Linnaeus' classification based on number of stamens.

Bonavia indicated that E.B. Cowwell mentions that the word 'nagaranga' is found in the oldest Sanskrit dictionary, *Amarakosha* written in 8th century A.D. It is not clear as to the type of orange referred to—Séville sour orange, the Portugal sweet orange or the suntara orange.

It is uncritically admitted that the Séville orange was the first orange to be introduced in Europe. The absence of sweet oranges immediately after the introduction of oranges in

Europe obviously indicated that the sweet orange did not exist at the source of introduction, otherwise this might have been introduced to Europe at the same time. De Candolle states that the sweet orange did not exist at that time as there was no available equivalent word in Sanskrit. A mention in Babar's *Memoirs* shows that 'santarah' was known in Bengal. An orange called 'nartan' in south India was known. It would be appropriate to conclude that the séville orange in south India was either indigenous or had been naturalized there from prehistoric times. Rumphius has mentioned Séville orange under the name of *Aurantium acidum.* It was unknown in Jawa, Baleya and Celebes and was uncommon in Malay Archipelago during those days.

There are two kinds of Portugal or Malta oranges : Bitcam and Seng-cam in China. It is most probable that the Portugal orange came to India from China via the Malay Archipelago. The pummelo are known as 'chakotra' and 'mahtabi'. The word 'chakotra' is derived from *chak* meaning a wheel denoting the gigantic size of pummelo. Whereas the 'mehtabi' is derived from the word *mehtab* which means the moon denoting a moon like size. The pummelo has retained its original name 'Batabi-lemboo' in West Bengal. The names chakotra and mehtabi possess a particular significance as they account for the old and new names of Jawa. The pummelo seem to be a recent introduction into India from Jawa.

The pummelo probably originated in southern China or Cochin China. A. de Candolle indicates that this has originated in the islands of the Pacific in the wild form. This has been placed as *C. decumana* which has many varieties growing in forests. It seems that the word 'pummelo' has been derived from the Dutch 'Pummelose'. Captain Shaddock was the first to introduce into West Indies and hence named 'Shaddock'. Bonavia (1890) mentions that it does not seem to be a distinct species but is an offshoot of the *C. aurantium sinense* Gall.

ORANGE

The suntara and kamla oranges are sweet but are quite different from Portugal oranges which have a close-fitting skin. 'Suntara' a Sanskrit word is a derivation from 'suntarati' means a thing which can float well. It is known that the

Suntara orange floats well. This occurred in the form of wild or supposed wild sweet oranges in Garhwal, Sikkim and Khasi Hills. This may be a supposed parent of the cultivated orange whether sweet or bitter. Similarly these have been reported by several wokers in the forests of the eastern Himalayas. This 'Sintra' is totally different from the orange of 'Cintra', a town of Portugal which is probably a corruption of the word 'Suntara'.

Its extensive cultivation from Kumaon to Khasi Hills and a little towards the coasts of India is indicative that the Portuguese probably did not introduce it in India. The orange exists in different names in Khasi. The most common are 'U soh niamtra', 'U soh syiem', 'U soh myngor', 'U soh kwit', 'U soh nairiang', 'U soh kymphor.' The word 'Syiem' in Khasi means 'Raja' or king. It appears that Khasis considered this to be the best orange. It is interesting to note that the generic name of oranges is 'U soh' in Khasi. In Malay Archipelago a modified i.e. 'aussi' and 'ussi' form of the same word is used as generic name of the citrus. These evidences indicate that either it came to Khasi hills and Malay Archipelago from a common source or that it was introduced from one place to the other.

It is established that 'Suntara' is completely distinct from the 'Cintara' orange in Portugal which could not have been introduced under that name by the Portuguese. Its cultivation from time immemorial in the north-east Himalayan region suggests that this must have been indigenous there or might have been introduced from the Malay Archipelago in ancient times.

The mandarin orange, *C. reticulata B'anco*, is a recent introduction to India. De Candolle indicated that mandarin oranges had their origin in Cochin China from there have been carried into different parts of the world. It is considered as a fancy fruit and as such commands a fancy price in season.

CITRON

Theophrastus, the Greek botanist mentions that the citron was in abundant in Media and Persia, three centuries before Christ. Its cultivation was well-known to Jews. The Jews used citron in their ritual in ancient times. It is believed that the citron might have been introduced into Greece and Italy from Persia and the warmer regions of Asia. Royal (1839) found the species *C. medica* growing wild in the forest of northern India and described India as the original home of citron. Hooker in *Flora indica* mentions that it is found wild in the foothills of the Himalayas. The species *C. medica* included the lime, lemon and citron and some other types. Watt (1908) indicated that the lemons reported growing wild were probably limes and included the 'jambhiri', 'kakhzi', 'khatta' etc. in the sub-species, *C. medica* var. acida. Bonavia (1890) stated that the popular Indian names for the citron, 'turanj' (from the Persian) and 'bajoura' (from Bajour in Afghanistan) indicate that some varieties if not the whole species entered India from the west. In the case of 'modhkakree' the 'madha' seems to be a sanskrit word meaning sweet and (kakree' may be melon and hence the large sweet citron of Almora is called 'madhkakree'. The citron are commonly known as 'mauling' and 'mapala' on the western coast of India. It seems probable that the citron originated in the area northwest of India.

It is claimed that the word Jambhiri has a Sanskrit origin. Dutta (1958) has reported that the jambhiri, *C. jambhiri*; Lush, is found in an array of different forms in Assam in a semi-wild state. This made him conclude jambhiri to be a native of this region. He has reported six distinct forms of jambhiri. It is known as 'soh-myndong' (*soh*=fruit, *myndong*=round) in Khasi hills, 'godha tulia' (*godha*=rough or warty) in Assam, 'jamburi' in Madhya Pradesh, 'kanda-narangai' in Madras and 'jathi khatti' in Punjab.

LEMON

In general lemon is considered as a variety of citron *C. medica* var. limonum. Although the foliage of lemon is different several intermediate forms are encountered. Little is known about the origin of the word lemon. It seems that it came to European languages through the Arabic word 'leimun'. Subsequently as the details of the species became available, it was called in Latin 'Limonium' and in pharmacopocias 'Limoneum'. The lemon is generally considered to be indigenous to India. However, Bonavia (1890) considers that it originated in Malaya.

LIME : *C. aurantifolia* Swingle

The 'kagzi nimboo' is commonly grown all over India, but it does not seem to be in the wild state. The lime is often, confused with the lemon obviously making it difficult to trace its history. Bonavia (1890) considers it to be of Malayan origin. The Malays call it 'lemon nipis'. The generic name for all citrus is called 'lemboo' in Bengal. Ample number of 'limboo' were found in Emperor Babar's time. The precise equivalent word is not traceable in Sanskrit.

4

Commercial Importance of Citrus Fruits

CITRUS fruits are grown all over India. These occupy about 0.1 million hectares of which Madhya Pradesh, Madras and Maharashtra are the largest. These are grown on a wide range of soils varying from shallow open soils to heavy black soils. The citrus flourishes in well drained alluvial or medium black soils of loamy nature. These have a wide range of adaptability and can adapt themselves from dry to heavy rainfall areas of the Cherrapunji region. The orange wood requires considerable amount of lime for a healthy growth. These flourish well in many parts of Meghalaya e.g. Shella area. The orange gardens are not irrigated in Khasi and Garo Hills. They do not water even young trees. These people depend entirely upon nature. About 750 trees are grown in a hectare. In Khasi and Garo Hills these gardens are commonly uncared for. The orange tree is simply stuck in the ground and left to nature. The producer does take a little care, by way of removing the competing weeds and cutting parasitic plants and dead wood once a year. The surface soil is perpetually renovated by the deposition of soils brought about with the floods. Presumably, the fallen leaves are another source for nourishing these gardens. These plants certainly require sufficient quantity of nitrogen, phosphoric acid and potash for the attainment of their proper development. The citrus fruits, in addition to these need lime, magnesia, sulphuric acid, chlorine and silica in appropriate quantities. Encouraged by the usefulness of citrus fruits particularly sour ones, the Khasi and Garo prefer to grow these anywhere in the vicinity including the backyard of their houses. At certain places the daily drainage of water from bathing etc. if led by channels to these trees would suffice for watering in summer. The lime and lemon could be easily grown from seed and layers.

Orange, lemon and lime are known for their taste and daily domestic uses. The relative proportion of sweet and sub-acid determines the taste of the orange. The ripe orange has a nice taste. However, there are acidless varieties of 'metha' orange. The sweet lemon among the lemons and 'madhkakree' among citrons occupy the favourite position. The sweet lemon normally fetches a better price than the oranges. The limes and lemons which mature in hot weather are invaluable. These make valuable drinks which are used in case of exhaustion. These are very useful for quenching thirst in the hot weather. These are also used to serve as cool drink of lemonade or orangeade to all types of people including sick persons. Almost all the parts of the citrus plants which include the branches, leaves, flowers and fruits are used in one or the other form. These have numerous domestic uses. The rind of the malta lemon is used to flavour a sweet dessert. The rind of the citron fruit candied is well-known as a delicate sweetmeat. These rinds make good comfits. The pulp is also preserved in sugar. In the western coast of India, there exist very large varieties. The thick sweet skin after peeling off the bitter rind is eaten. A preserve is made of the thick skin of the citron in North India. The juice of the sour lime is commonly sprinkled on soup or curry to impart a pleasant agreeable flavour and taste. The pickle in its own juice is prepared. It is an effective medicine for indigestion brought on by overeating. The rind of the sour lime is scraped a little to thin it. This is steeped in juice obtained from other fruits of sour lime, after adding common salt. These are exposed to the sun for a few days. This has a pleasant taste and is a useful medicine for indigestion. These are used for preparing Sherbets and lime juice. The fruit is also pickled. The galgal contains a good amount of juice. The lime and lemons are useful sources for manufacturing citric acid which is used for pharmaceutical purposes and for aerated waters. Citric acid which has considerable demand might be manufactured from all sour oranges or from any other sour citrus. The Séville orange is most useful for making marmalade and preserve. This Séville or bitter orange has a very aromatic and bitter rind and sour pulp. Lemon jelly is another delicious preparation. Any fresh acid citrus fruit such as Séville orange, the lime could be used for preparing jelly but the best jelly is that made from the fresh Malta lemon. The sugar used for preparing jelly must be of white crystalline kind otherwise the slightest amount of treacle in the sugar will make a jam and prevent the jelly from setting. The citrus rind can be made into candied peel for purposes of confectionery. A combination of citric acid flavoured with lemon oil enter into the composition of lemon drops. The essential oil of the rind loses its attractive aroma if kept over a period of time. Freshly made essential oils are required for perparing quality lemon drops. Similarly fresh peel is relatively soft than the imported peel. It is therefore desired to use fresh peel for quality standards. The proportion of acid pulp to rind can always be regulated according to individual taste. Citron peel is preserved in syrup and is called 'turunj ka muraba' in places like Lucknow. The citrus peel which is not too bitter can also be candied. The juice of sweet oranges is used for preparing wine. It is believed that this beverage is manufactured in considerable quantities from sweet oranges. Séville orange is used for cleansing copper wares and at times for washing the hair. It is used as an antidote against land leeches.

Perfumery

The leaves, flowers and fruits are used for perfumery. The syrup of the orange flower is mostly used as a flavouring agent and to perfume external applications. Two distinct perfumes are procurable from the orange flavours. The type of fragrance is determined by the method adopted for procuring them. Different types of scents are made from the orange flowers. The Portugal or sweet orange (C. aurantium) is the most useful source for the extraction of the finest otto. The next point is that from the flowers of the Séville orange. The leaves and young unripe fruit of several varieties of citrus are also used in preparing otto or perfume.

Oils

Orange oils are also used for scenting soap. The rind of lemon oil is rasped and subjected to expression or is distilled and yields an essential oil known as 'essence of lemon' or 'citron-zeste' according to the method adopted. Lemon oil is of faint yellow colour and bitterish aromatic taste. The oil dissolves sparingly in rectified spirit. The chief constituent of essential oil is the terpene, hydrocarbon and a small quantity of cymene and of a compound acetic ether.

Medicine

Lemon oil is a stimulant and carminative when given internally as medicine and stimulant and unpleasant when applied externally. Bargamot oil extracted from the bargamot orange (*C. aurantium* var Bargamia) is of green tinct, clear limpid liquid, bitterish, of very fragrant odour and aromatic taste. It has the property of being miscible with oil of turpentine, glacial acetic acid and rectified spirit. It is used for giving an agreeable odour to ointments and other external applications. It is also used in perfumery though rarely in medicine.

The lemon is made into marmalade and antiscorbutic. It is made into a preserve which is used for dysentery. Citron rind is hot, dry and tonic; the pulp cooling and dry ; leaves, flowers and seeds are hot and dry. The juice is refrigerant and astringent. Orange flower water is used as a lotion for eyes, skin and hands. It is used in confectionery and for mixing with nauseous medicines. It is used as a remedy for scurvy and also dysentery, diarrhoea and acute rheumatism. It forms an excellent drink in fever and inflammatory infections. It is considered as an antidote to animal poisons. The lemon juice speedily becomes unfit for medicinal use as it is liable to spontaneous decomposition. It should therefore be preserved properly. Relief is derived by people suffering from acne by rubbing the fresh rind of oranges on their faces. Relief is also obtained by people suffering from eczema if the rind mixed with a little water is rubbed on the affected part.

Jambhiri is used to quench thirst and as an anti-emetic. Lime is used in food, medicine and the arts. Lemon juice along with the common salt is recommended as a remedy for dysentery, dry belly-ache and putrid sore throat. It corrects foetid breath (Drury). Slices of lemon provide a useful application for sores including scorbutics.

'Tincture of orange peel' is prepared by extracting the active principle of orange peel by means of alcohol. It is primarily used for disguising the unpleasant flavours of medicines. The extract of the rind of Séville orange is used as tonic and stomachic. It is used in combinations in preparing a tonic for dyspepsia and general debility. The orange peel is fruitfully employed in the form of infusion, tincture or syrup. The leaves of the pummelo are useful in epilepsy and convulsive cough. The bargamot orange (C. aurantium var Bargamia) provides refrigerant drink in meals, scarlatina and other fevers. Fresh lime is useful in relieving bites, irritation and swelling caused by mosquito bites.

The decoction of lemon and of the 'khatta' orange has given useful results to recover from cochexia left by intermittent fever. The decoctions are the best splentifuges in cases of recently enlarged spleen. It is very useful for improving the appetite and digestion and diffusing vigour through the muscular systems.

Timber

The orange trees generally yield white and yellowish white moderately hard, close and even grained wood. It is used for preparing handles of implements.

Gum

The orange trees yields scanty unimportant gum.

5

Morphology and Classification of Citrus Fruits and their Relatives

THE CLASSIFICATION of citrus fruits is complex and this difficulty has tremendously increased by the numerous attempts made to classify the citrus fruits from time to time. In fact these classifications instead of simplifying the problems have only made them more complicated. Citrus fruits show colossal changes in their features by change of climate, soil and culture and if grown from seeds, under different circumstances further changes might be expected. The extent of sweetness, juiciness, smoothness, tenderness of skin and the size of the fruit are some of the characters which are liable to change with the change of environment and hence it is of utmost importance that the characters of the tree, flower, fruit and sweetness should be considered in grouping them. The classification of scientists was based not only on the available characters; they also took into account the several characters which these plants had thousands of years ago. This made more intricate the complicated problem of classification. Although the classification is based on the group of characteristics, even then it is not always easy to place a tree in its proper species. There is no classification of citrus fruits which has found general acceptance. In principle, the original species must be described from the truly wild specimen. Botanists have deviated from this general rule, else they might have to eliminate the entire *Citrus* genus, for even the generally accepted species are found only in cultivation. This further complicated the classification e g. grape fruit and lemons. If the wild forms of the available citrus fruits should be found they would perhaps differ in some degree from the cultivated ones. The chances of difference between the cultivated and wild ones are more if these are farther away in time. When these plants were brought into cultivation such variants as arose in different ways attracted the attention of the grower and the same was perpetuated by one way or another. This might have resulted in the removal of some of the cultivated citrus fruits far beyond from their wild progenitors. As there is no law against establishing a species in cultivated form, nothing else could be done than to use the material in hand.

By doing so, we will have to disregard for the time being the possible existence of the older or more primitive types, from which the present citrus have been derived. In case the older forms should be described they may be entitled to specific ranks. Walter T. Swingle studied interesting collections of living plants in the green house of the United States Department of Agriculture at Washington and came forth with his classic work of grouping under the family Rutaceae. While classifying he separated the Kumquats as species and placed them in the new genus *Fortunella* named after the English plant explorer Robert Fortune and the re-establishment of botanical name *Poncirus trifoliate* Refinesque for the trifoliate orange. Subsequently, Swingle made some changes in nomenclature based upon further studies; he grouped grape fruit as a separate entity from Shaddock. The retentions of *Fortunella crassifolia* and *Citrus mitis* as species and not hybrids as Fortune placed them in 1946 are the noteworthy changes.

All the citrus fruits were earlier deemed to be the members of the genus *Citrus* but presently the citrus fruits are referred to three genera—*Citrus*, *Fortunella* and *Poncirus*. The trifoliate orange formerly called *Citrus trifoliate* is now known as *Poncirus trifoliate*. The Kumquats comprise the other new genus, *Fortunella* and were earlier called as *Citrus japonica*. Presently the oval Kumquat is classified as *F. margarita* and the round Kumquat as *F. japonica*.

RUTECEAE

Trees or shrubs, rarely herb plants are mostly native of tropical and sub-tropical countries primarily in the Old World. Leaves usually alternate, compound, (Palmately or pinnately) aromatic, 1-foliate, serrated, sometimes reduced to spines, dotted with translucent glands containing an essential oil, petiole often winged. The leaf in general is simply an expansion of the bark. The leaf of the citrus fruits is studded with two or more sets of glands viz. large glands at certain distances, the intermediate are filled with secondary and tertiary oil cells. The margins of the leaf are crenated. The big crenations have smaller crenations. The angle formed by both these crenations have oil cells. The crenations are formed only towards the tip half of the leaf in several oranges whereas these generally commence close to the tip in the case of citron, lemon and other allied types. Rarely a variety may be encountered without crenations but the oil cells at the edges are present. In *Citrus*, two or more buds are produced from the normal axillae of the leaf, these either form fully developed branches or remain abortive and form spines.

Flowers

The flowers are transformed branches either coming out of the axilla of an ordinary leaf or from that of an abortive leaf usually called bract. Bisexual, (sometimes unisexual and the plants then dioecious in some Zanthoxylum), generally, regular hypogynous or perigynous, usually axillary, solitary, fascicled or in small cymes, white or pinkish, sweet scented; sepals : 3-5 or lacking, cupular or urceolate; petals : 3-5, rarely none, usually distinct, thick imbricate, the disc present varying from an apparent receptacular swelling to cup like structure; stamens : 3-10 or more basically in 2-whorls; stamens, those of outer ones usually opposite the petals, stamens as many or twice as the petals, (in *Citrus*; *Aegle* more), separate or united inserted on the receptacle. All stamens attached at the base or rim of the base, some occasionally reduced to staminodes, distinct or rarely basally connate, rarely adnate to petals, usually straight and unequal, anthers 2-celled introse, dehiscing longitudinally, opening along the inner face, generally versatile; ovary : usually 4-5 locular, multi-locular in *Citrus* with axile placentations, ovules, 2 (rarely 1) in each locule ; style : stout, deciduous; fruits: the citrus fruits probably have originated in two whorls of carpels, the outer is rind whorl and the inner is pulp whorl ; a berry with

B. 1 : Poncirus trifoliate

B. 2 : Citrus limon

B. 3 : Citrus jambhiri

B. 4 : Citrus indica

removable rind and separable sections, packed fully of pulpy tissue which is composed of large cells and developed from the endocarp, projecting towards the centre thus embracing and surrounding seeds; capsule, drupe or hesperidium; Seeds : variable in number in each cell, pendulous or horizontal, tasta membranous; embryo sometimes 2 or more in one seed; plano-convex cotyledons.

The family comprises many members of economic importance. The most important among them are citrus fruits, orange, lime, lemon and citron and Kumquat (Fortunella), ornamental include hop tree (Ptelea), prickly-ash (Zanthoxylum), cork tree (Phellodendrom), common rue (Ruta), trifoliate orange (Poncirus) and orange jessamine (Murraya).

The family has about 140 genera and 1300 species. The following three genera are of special significance in the present studies.

Leaves of 3 leaflets deciduous—*Poncirus*.
Leaves usually 1 leaflet (apparently simple) persistent.
Ovary 3-5 or 6 celled—*Fortunella*
Ovary 8-celled or more—*Citrus*

Poncirus

(French : *poncire*, a kind of citron). Small spiny tree with deciduous trifoliate leaves; Petals : narrowed and claw like at the base; opening wide and flat; Stamens : numerous, free; Ovary : 6-8 celled; Flower : buds produced in summer protected by bud scale, rarely sessile, borne on the previous season growth; Fruit : pubescent, juice vesicles with hair like attachments containing an oily, sticky aromatic fluid; Seeds : many with white cotyledons.

Poncirus trifoliate

Small spiny tree with an inedible fruit. It is known for hardiness and hence is used as root stock particularly in regions of severe frost. It is rarely found in India. This seems to be indigenous to Central or Northern China (Illustration B1).

Fortunella

Kumquat—small evergreen trees or shrubs densely branched, spines small and axillary or absent; Leaves: blunt-pointed, unifoliate, dark green above, paler beneath, thick, prominently dotted with glands on the lower surface; Petiole: normally veined; Flowers : white, commonly smaller than those of flowers of citrus, borne single in pairs of clusters in the axils of the leaves or new shoots or on the shoots of previous years of growth; pedencles bracted with one or two small bracts; Petals : 4-5 nearly oblong; Stamens : 16-20, varied in length; filaments united at the rings except at the tips; Ovary : generally 3-6 celled, 2 ovules in each cell; Fruit : small, rind thick, sweet edible and small. It is native in South-East China. *Fortunella japonica* Swingle round kumquat : shrub densely branched, small sharp spines; Leaves : small, blunt pointed tapering towards base; Flowers : white small axillary; Fruit : small, round, orange coloured juice and acid with sweet and edible rind; Seeds : 3-6 small, these are of little commercial value, but are grown for ornaments. The 'Hazara', so called Kumquat of Northern India is similar to true Kumquat but is the calamondim. The calamondim is believed to be native of Philippines. It has been classified as *Citrus macrotera* (Mitis).

Swingle (1943) considered the calamondim to be the hybrid between the genera *Citrus* and *Fortunella.* Tanaka, however considers it *C. madhurensis*.

Orange, lemon, lime and citron have been critically dealt with by a large number of workers. Brandis (1874) in his *Forest Flora* presented a concise and pregnant account of the available Indian *Citrus* species. Risso and Poiteau's (1811) in *Histoire Naturelle des orangers* published monumental pieces of interesting work which dealt with the description of the species. A number of wild species have been described by Hooker (1872). De Candolle (1866) in *Origin of Cultivated Plants* has endeavoured to fix the limits or regions to which the various *Citrus* species may be viewed as indigenous. While doing so he has given due consideration to the extensive studies of early literature and philological work. There has been considerable difficulty in grouping the species of the genus partially because of the similarities and partially because of hybridization. Many of the characteristics on which classification is based are not constant within the species. George Watt (1889) mentions that the genus *Citrus* has three important species viz. *Citrus medica* Linn, *C. aurantium* Linn and *C. decumana* Linn. Sir J.D. Hooker in his *Flora* includes four varieties under *C. medica* and three varieties under *C. aurantium* and includes *C. hystrix*, an additional species. Kruz in *Flora of British Burma* divides one of the varieties of *C. medica* into two types, *C. limetta* Risso as synonym under *C. nobilis* Lour (the mandarin)—a species which he regards as quite distinct from *C. medica* Linn.

Watts (1889) has drawn the following analysis from the work of Hooker with two additions from Kruz.

Young shoots and leaves perfectly glabrous; transverse vesicles of the pulp concrete.

A Shrub

Young shoot purple, petiole more or less naked; Petal : generally tinged with red ; Flowers : often unisexual ; Stamen : 20-40 ; Style long thick ; Fruit : globose, ovoid or oblong, often mammillate, rind very thick and rough—*C. Medica.*

A tree 15-25 feet in height ; Petiole short winged ; Flowers : small, white, usually solitary; Style : long, thick ; Fruit : globose or somewhat oblong, not mammillate; rind very thin, nearly smooth, shining, yellow or orange coloured—*C. nobilis* and (C. limetta).

A small shrub : leaflet smaller than the broadly winged petiole; Flowers : as *C. nobilis*, only pedicillate and clustered in the axils of the leaves ; Style : very short ; Fruit : globose or ovoid a little larger than the size of walnut; rind : thick yellow—*C. hytrix.*

A tree, young shoot whitish; Petals : more than twice the length of those in the two preceding species ; Flowers : bisexual ; Stamens : 20-30; Style : long, thick ; Fruit : globose or flattened, pulp sweet acid or bitter—*C. aurantium.*

Young shoots and under surface of leaves pubescent, transverse vesicles of pulp distinct—*C. decumana.*

There are a number of characteristics which determine the grouping, the essential characteristics to be taken into account are, the extent to which the petiole is winged; the colour of the flowers : pure white for the oranges and pinkish white for lemons ; shape of the fruit : globular, non-mammillate in oranges, pear-shaped or mammillate in lemons. Intermixing rendered through hybridization is not an uncommon process. This obviously complicates the process of classification. Swingle indicates that many of these are chance hybrids or mutants : cross-pollination by insects is another cause for the production of hybrids. It is doubted whether some of the well-known species were actually hybrids. Swingle prefers to use the term 'satellite species' for such doubtful cases. He has divided the genus citrus into two sub-genus *Eucitrus* constituting the important edible species, *Pepeda* having pulp vesicles with droplets of acrid oils which makes the fruits unpleasant. In addition these have notable distinct characters of the leaves and flowers.

CLASSIFICATION OF CITRUS

Walter T. Swingle (1943) recognized 16 citrus species whereas Tyozaburo Tanaka (1954) concludes that there are 140 species.

Citrons

C. medica Linn, 2n=18. Considered to be a native of India (De Candolle 1886, 1912), wild citron has been encountered in Khasi and Garo Hills and Chittagong. Royle (1839) observed this species in the wild form in the forests of north India. Hooker (1872) indicated the presence of citron along the foothills of the Himalayas from Garhwal to Sikkim, Khasi and Garo Hills, Satpura Range in Madhya Pradesh and the Western Ghats. However Bonavia (1890), Swingle (1946) are of the opinion that the citron (*C. medica*) is not of Indian orgin. Lowther (1857) encountered citron plants in the wild state. The citron is called 'Bira Jora' or 'Bokel-khowa-tengu' in Assam plains, 'lume-han-thor' in Mikir Hills; 'haijange' in Manipur, 'sohmadheh' or 'soh menong' in Khasi Hills, 'sutrung' in West Bengal, 'mokari' in Punjab, 'madhkakree' in Uttar Pradesh and 'mokri' in Madras. The fruit is known as the apple of Media. Theophrastus called it Median or Persian apple. Its name, *C. medica* is derived from the country Media.

The citron is believed to have been introduced into Italy and Greece from Persia and warmer regions of Asia. Gradually it has spread all over the world including the cold regions. The citron did not draw the attention bestowed on the orange.

This was due to the reason that the important product obtained from the citron was candied peel, a commodity of limited use to the pioneers.

The trees have short, thick, straggling thorny branches. The leaves are more or less oblong with serrated margins and the petioles are small and wingless. The flowers are large and are mostly curved before opening. These have deep purple pigmentation, style is thick. The fruit is large or obovoid, usually mammillate, rough or irregular or warted, yellow. When ripe, the pulp is pale yellow, scanty, mildly acid or sweetish, insipid. The rind is rough, and thick and is good for making pickles. Broadly citrons are of two types, sweet and acid (Fig. 1).

Lemon : *C. limon* Linn, Brum f. syn *C. medica* var limonum of Watt. 2n=18, 36

This name has been accepted for true sour lemons without any contradiction. Glidden (1937) studied the history of lemons in detail and concluded it to be native of the Eastern Himalayan region. A number of lemons both wild and cultivated have been reported in this region. The lemon alone constitutes an important group consisting of ten distinct varieties (Dutta 1958). Lowther (1857) reported a number of high quality lemons growing in a wild state throughout the forests of Assam.

It has been reported by some workers that south-eastern Asia is the natural home of the lemon. Laufer (1934) mentioned that lemon is not indigenous to China but is an introduction in China as the word 'li-mung' used in the earliest Chinese literature for lemon is altogether foreign to the Chinese language. The lemon was perhaps introduced into Europe about the same time as the sweet orange. The lemon is called by various names in India, limbu, limu, nimbu (Hindi) etc. Glidden (1937) who studied the historical records and languages for equivalent words of lemon concluded that lemons are natives of the Eastern Himalayas (Illustration B2).

The lemon is a small tree about 10-12 feet high, with thorny branches; Leaves : slightly winged or wingless, petiole; Fruit: 3-5 inches long; Segments, 9-12; oblong or ovoid fruit with nipple shaped extremity, bright yellow, pale yellow, acid pulp. It is grown throughout India including the courtyards of homes and in varied sizes of orchards. It is adaptable to a variety of soils and elevations. There are a number of similarities in lemons and citrons. These have a similar pattern of development and growth, purplish flowers and similar leaves except that they are generally articulated in lemon. However, they have different characteristics in the fruits. The fruits of the lemons are relatively smaller with thin rind and plentiful acid juice. The lime and lemon which were earlier included with citron in species. *C. medica* have now been recognized as a separate species.

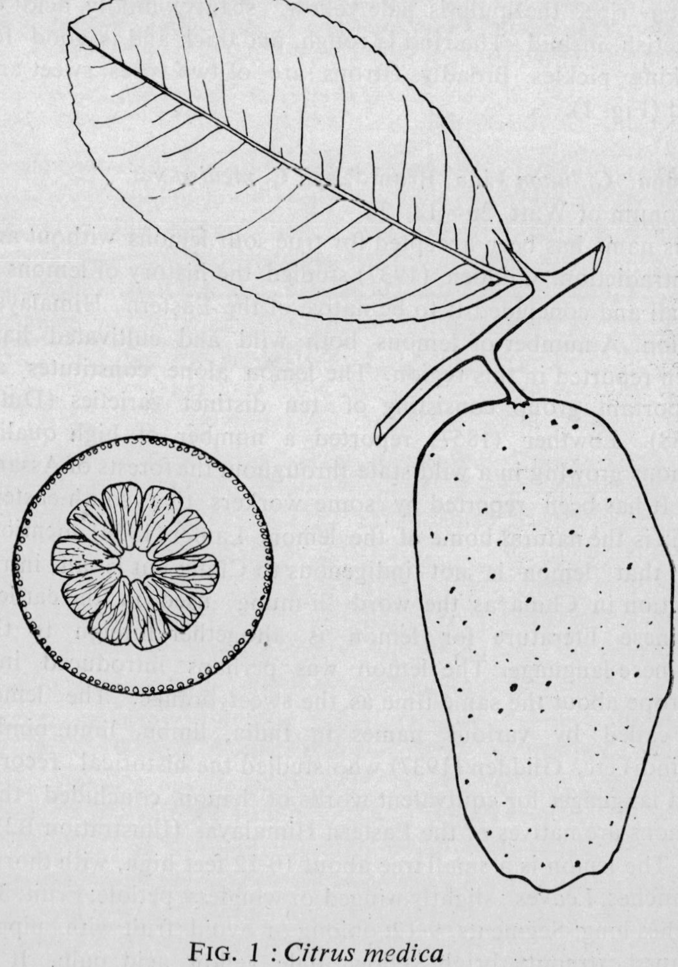

FIG. 1 : *Citrus medica*

The lemons have been recognised now as *C. limon*. However Swingle considers this species as 'satellite' of the citron as he believes that it is a hybrid of citron and lime. Bonavia (1890) separated the true lemons from the hybrids resulting from the intermixing of citron and lemon and placed them under *C. medica* var. bajoura. There are several hybrids which are prevalent in the Eastern Himalayan region, some of these are called as 'Soh-mandrit', 'jora-tenga', etc. Webber (1946) attempted to divide the lemons into two distinct groups based on sweetness or acidity of the pulp-vesicles, later when Risso (1818) designated sweet lemons as *C. lumia*, he took the flowering and fruiting habits into account (Fig. 2).

Sweet lime *C. limetta* Risso

The sweet lime is commonly cultivated in different parts of India. It is a heavy yielder and fruits are available later than the other citrus fruits. The fruit is mildly sweet with practically no acidity. It is sometimes included with true lime. A closely resembling fruit in Europe is called *C. limetta* Risso. However, Tanaka (1937a) is of the opinion that the Indian sweet lime is different and placed it as *C. limettoides*. Webber (1946) realised the difference between sour and sweet lime but preferred to put it as a variety under the species *C. aurantifolia*. He, however, doubted whether the sweet lime was a hybrid. Roxburgh (1832) called it as *C. acida*, while Hooker (1872) recognized it as a variety and hence placed it as *C. medica* var. limetta. Lushington (1910) recognized it as *C. limetta*. However the sweet lime was ignored by Swingle.

It has been reported by Wright to be growing wild in the Nilgiris and this made him consider it to be indigenous to India. This has been mentioned as 'amrat-phal' in the *Memoirs* of Babar. It is also possible that it was introduced in Assam during the middle of the nineteenth century. It is commonly known as 'Soh-bathiang' (Soh=fruit, bathiang=sweet) Khasi Hills; Muro-muri (Mau=honey) Assam; Mitha nebu=Hindi, Kolumichangai—Tamil Gajanimma—Telugu, imbe—Kannada. It has medium sized trees with relatively thick branches; leaf : practically no petiolar winged, prominent oil gland, light green, distinct aroma; Flowers: pure white without tinge; Stamens : many. Fruit : globose, ovoid, segments 10-11; smooth, relatively

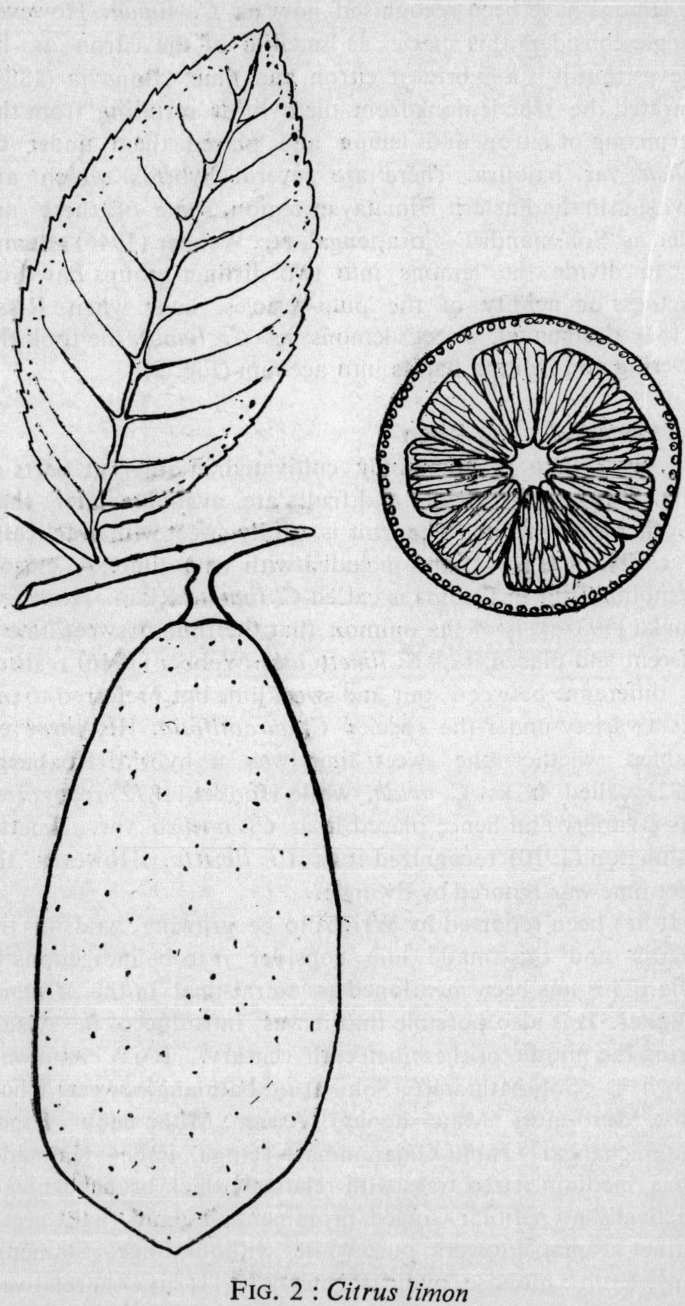

FIG. 2 : *Citrus limon*

bigger than the sour lime, so do the seeds, the rind is relatively thick, juice sweet with sweet pulp vesicles. It is usually the source of bud material for mandarin orange (Fig. 3).

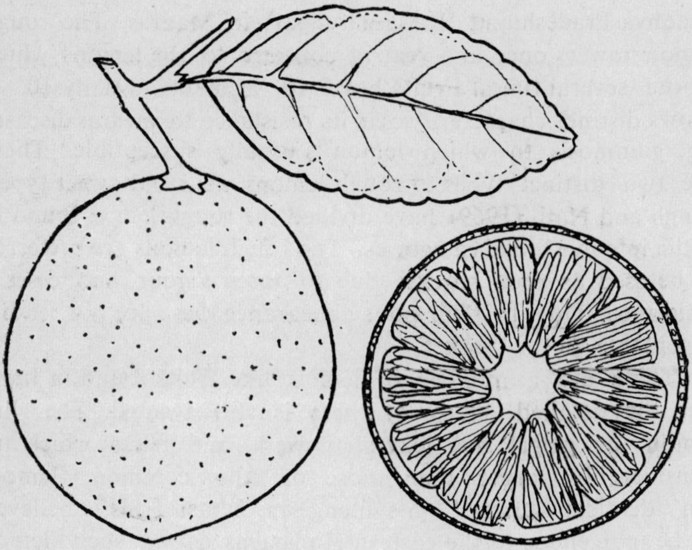

FIG. 3 : *Citrus limetta*

Citrus jambhiri (Lush)
This has been recognized as separate species by Lushington (1910) and later by Tanaka, however, Swingle does not recognize it as a separate species.

It is found in semi-wild state in the north-eastern Himalayan region. Several distinct types of jambhiri exist in India. Bonavia (1890) indicated that jambhiri is not known outside India. It is very similar to the rough-lemon of Florida. Swingle (1946) indicated that either it is indigenous to India or it has been widely naturalized. There have been different opinions regarding the relationship of the true lemon and jambhiri or rough lemon. Hays (1945) indicates that it might be a hybrid or sport of the other. The rough lemon or jambhiri has been referred to as *C. limonia* Osbeck and by several workers (Cheema, 1934; Naik, 1940; Nandi *et al.* 1943). Lushington

(1910) recognized the rough lemon as a separate species, *C. jambhiri* (Illustration B3).

The different forms of the rough lemon are known as 'Soh-myndong' (Soh=fruit and myndong=round) in Khasi Hills, 'Gudha Tulia' in Assam, 'Jathi-Khatti' in Punjab, 'Jamburi' in Madhya Pradesh and 'Kada-narangai' in Madras. The rough lemon flowers once in a year in contrast to the lemons which bloom several times. Fruits has 8-10 segments, usually 10. It shows distinct characteristics in its resistance to various diseases e.g. gummosis to which lemon is usually susceptible. There are two distinct types of rough lemons, acid and sweet types. Singh and Nath (1969) have divided the rough lemon found in India into five broad categories. The rough lemons are preferred to be used as root stocks due to their vigour and disease resistance. Due to their poor appearance these are not readily acceptable in the market.

There are a number of lemon like fruits found in India but the detailed study of these is still awaited. The hill lemon or galgal has leaves, flowers and fruits which are considerably larger than those of the common lemon. An identical variety 'Soh-synteng' in Khasi Hills is believed to be indigenous to the eastern Himalayas as no such lemon has been reported from elsewhere. The hill lemon can relatively withstand better both heat and cold than the common lemon. The hill lemon bears only one crop in a year in contrast to the common lemon which flowers throughout the year, in addition these are heavy yielders with excellent quality fruits (Fig. 4).

C. karna Raf.

It is commonly known as 'karna', 'khatta nubu', 'id nimbu' or 'kharna khatta'. Although this fruit is frequently referred to in this country as citron which it in no way resembles. This fruit does not have an English name. It is commonly used as rootstock because of hardiness. It is commonly called as 'Soh sarkar' in Khasi Hills (Soh=fruit, sarkar=government). The name indicates that it was introduced there by the Government. It is called as 'kharna-khatta', 'Mokari' (Punjab), 'id lemon' (Maharashtra). It is probably indigenous to India. The species has very low polyembryony resulting in great variation in the fruit's shape and surface. The leaves and flowers of this species are similar to those of the rough lemon but larger in size. It is a medium-tall tree, 15-20 feet, spreading, thorny leaf medium sized, narrowly petiolar winged, dark green, articulated margin, serrate; Flowers : both bisexual and staminate, bisexual more than the staminate,

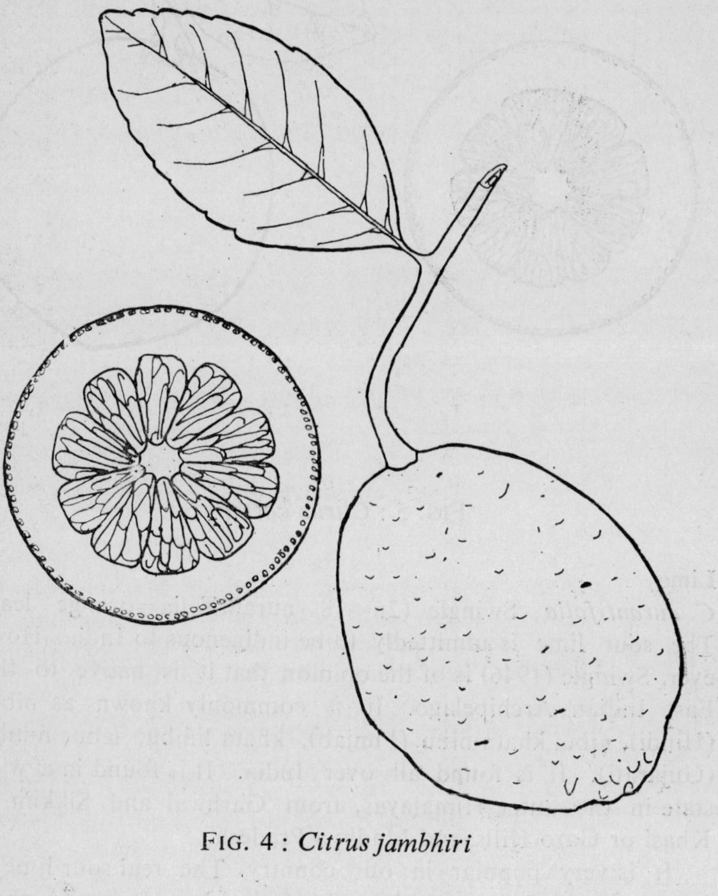

FIG. 4 : *Citrus jambhiri*

cymes, solitary and clustered; Petals: 4-5, purple outside, white inside; Stamens : many filaments joined at the base, top free white; Ovary : ovate, stigma capitate, Style greenish white; Fruit; large, shape variable, usually ovate oblong, yellow rind thick, Segments 9-13; adherence medium, oil glands present, mesocarp generally white spongy sweet; Seed : cotyledon white (Fig. 5).

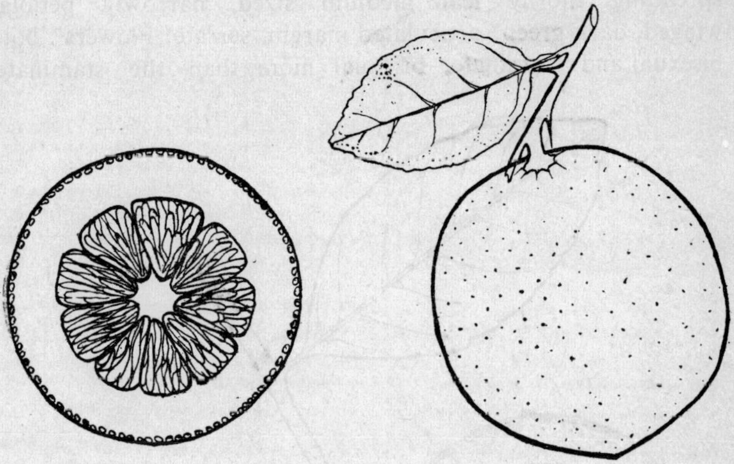

FIG. 5 : *Citrus karna*

Lime

C. aurantifolia Swingle (2n=18; aurantifolia=orange leaf) The sour lime is admittedly to be indigenous to India. However, Swingle (1946) is of the opinion that it is native to the East Indian Archipelago. It is commonly known as nibu, (Hindi), nibu, khata-nibu (Punjab), khata limbu; lebu, nimbu (Gujarati). It is found all over India. It is found in a wild state in the outer Himalayas, from Garhwal and Sikkim to Khasi or Garo Hills and Madhya Pradesh.

It is very popular in our country. The real sour lime is known as kaghzi everywhere in India. It occurs as a small dense branched forming small tree, 6-10 feet; branch, slender, drooping, thorny, small sharp; Leaf : small with prominent wing, elliptical to oblong winged petiole and articulated; Flowers: small, Sepals; greenish white; Petals: pure white, oval shaped; Style : purple; Fruit : round or oval, greenish yellow, smooth surface, prominent oil glands, thin rind, segments, 9-10;

greenish pulp aromatic. It flowers generally all the year round, mainly February-March. In India, lime is more popular as compared to lemon (Fig. 6).

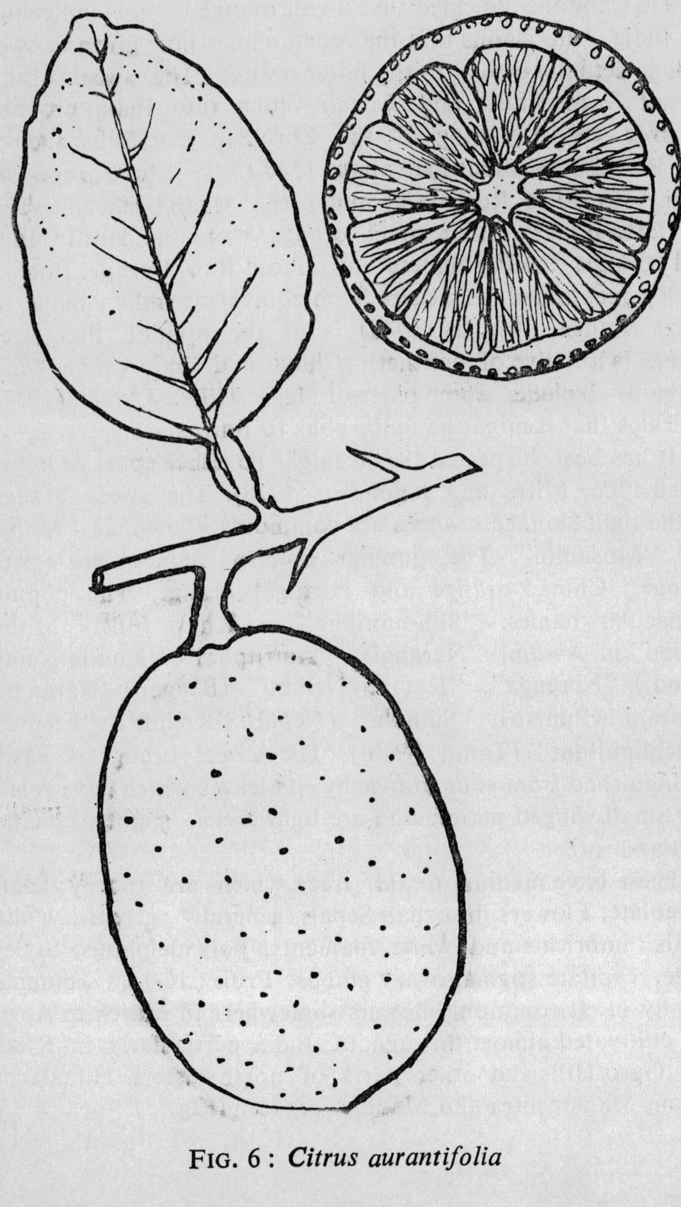

FIG. 6 : *Citrus aurantifolia*

Sweet orange

C. sinensis (Linn) Osbeck Syn. (*C. limetta* Risso; *C. aurantium* var aurantium proper, race 2nd and 3rd of Watt); 2n=18 (27,36)

De Candolle indicated that sweet orange is not indigenous to India and claims that the Asian names now given to sweet orange actually belong to the bitter orange. The sweet orange spread from China and Cochin-China into India probably towards the beginning of the Christian era. Subsequently, the Portuguese somewhere about 1548 took it to Europe and then it spread to Rome and along the Mediterranean. Vasco da Gama indicated that sweet orange was abundant in India and perhaps the Portuguese introduced it to Europe. Bonavia (1890) mentioned that there is no equivalent Indian name for sweet orange. Swingle (1946) is of the opinion that sweet orange is a native of southern China and Indo-China which probably includes some parts of Naga Hills. Tanaka (1937a) indicates that it might be indigenous to Burma.

It has been suspected that it might be either sport or hybrid (Basu 1906, Mitra and Khongwir 1928). The sweet oranges or the tight skinned oranges are commonly known as "Malta" and "Mosambi". The popular English names are sweet orange, China orange and Portugal orange. The popular vernacular names, "Soh-nairiang" in Khasi hills "Mitha-chakla" in Assam; "Narangi", "Amritphal", "Kumla Nebu" (Hindi); "Nerenga", "Kamla Nembu" (Bengali); 'Narangi", "Naranj" (Punjab); "Suntala" (Nepal); "Kichili", "Chebu", "Kichlipullum" (Tamil Nadu). The sweet orange is easily distinguished from sour orange by its leaves which have relatively small winged petiole and are light green, slightly smaller fruit and sweet.

These have medium to tall trees which are thorny; Leaf : lenceolate; Flowers: bisexual; Sepals: generally, greenish white, Petals : imbricate and white filaments, polyadelphous; Style : white, capitate stigma, ovary globos; Fruit : 10-13, segments, usually 11. It commonly flowers somewhere in March to April. It is cultivated almost throughout India particularly in Khasi and Garo Hills and other parts of north-eastern Himalayan region, Maharashtra and Madhya Pradesh (Fig. 7).

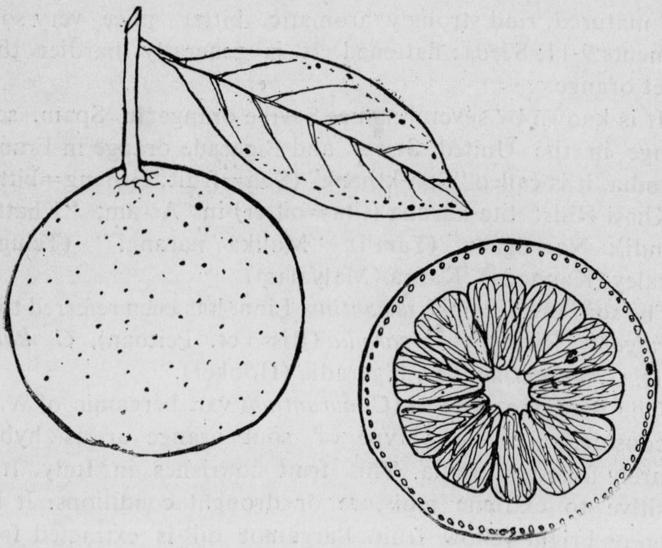

FIG. 7 : *Citrus sinensis*

Sour or Séville Orange: *C. aurantium* Linn (2n=18)
This has been reported to exist in India from ancient times and is considered to be of Indian origin. Webber (1946) indicated that this fruit was taken by the Arabs from India in about A.D. 922. Hooker (1872) emphatically indicated that it was found growing in India. However, its existence in India was not known to some of the earlier explorers of the Indian flora. There are references in literature to justify it to be of Indian origin. Bonavia (1890) mentions that "Kirnah" referred in the memoirs of Babar (1519) was a type of bitter orange of India.

It is a medium-tall tree with dense compact round head, densely foliaged thorns, alternate, long sharp pointed; Leaf : ovate pointed, articulated, alternate, petiole broadly winged, aromatic, characteristic peculiar scent of sour orange; Flowers : white sweet scented, somewhat larger than those of sweet orange (*C. sinensis*), axillary cymes; Calyx : cupped, segments blunts; Petals: prominently dotted with oil cells; Stamens : many, filaments united in group; Pistil : club shaped, smooth, ovary 10-12 loculed; Fruit : globose to obovoid, smooth surface, orange coloured or reddish when well matured, rind strongly aromatic, bitter, juice very sour, segments 9-11; Seeds: flattened. It is generally hardier than sweet orange.

It is known by séveral names Séville orange in Spain, sour orange in the United States and Bigarade orange in France. In India, it is called "Soh-kthang" (Soh=fruit, Kthang=bitter) in Khasi Hills; tita-karuna (tita=bitter) in Assam; "Khatta'; (Hindi); Narangans (Tamil); "Malika narangi" (Telugu), Haraley (Kannada), Karna (Malyalam).

The sour orange or *C. aurantium* Linn. has been referred to as *C. vulgaris* (Risso), *C. bigarodia* (Risso et. Peitoan), *C. karna* (Raf), *C. aurantium* var. bigaradia (Hooker).

The Bargamot orange (*C. aurantium* var. bergamic of Watt) is considered to be a type of sour orange or as hybrid is rarely found in India. This fruit flourishes in Italy. It is sensitive to extreme moisture or drought conditions. It has green or bright yellow fruit. Bargamot oil is extracted from its peel.

Kachili or Vadlapudi orange of Madras resembles Mandarin in having loose and rough rind. The pulp is sweet when ripe. It is included under sour orange which it resembles in foliage. It has been assigned a separate species by Tanaka, *C. mederas* Patana (Tanaka) (Fig. 8).

The loose skinned Mandarin orange *C. reticulata* Blanco (2n=18) The cultivation of loose skinned commonly called mandarin orange is rapidly extending in the Khasi Hills. It is unequivocally admitted that Cochin-China the southernmost region of the Indo-Chinese peninsula is the native place of mandarin oranges. Bonavia (1890) mentioned that the true mandarin while found in Ceylon did not exist in India. However, this orange is reported now growing throughout India particularly in Meghalaya, Sikkim, Assam, Madhya Pradesh, Maharashtra, Punjab, Madras and Mysore. The oranges grown in high altitude are all deeply coloured and intensely flavoured. The same type grown in different regions has yields varying in quality. In general, these are highly scented flat oranges. These have a peculiar smell common to both fruit and leaves. Various

B. 5 : Citrus grandis

B. 6 : Citrus grandis

B. 7 : Citrus macroptera

regional names have been attached to the mandarin orange e.g. the mandarin orange growing in Khasi Hills is called Khasi-mandarin. It is called "Soh-niamtra" in Khasi Hills; "Narang" in Garo Hills; "Kamala" in Bengal, "Sum-thira" in Assam; "Santara" in Hindi.

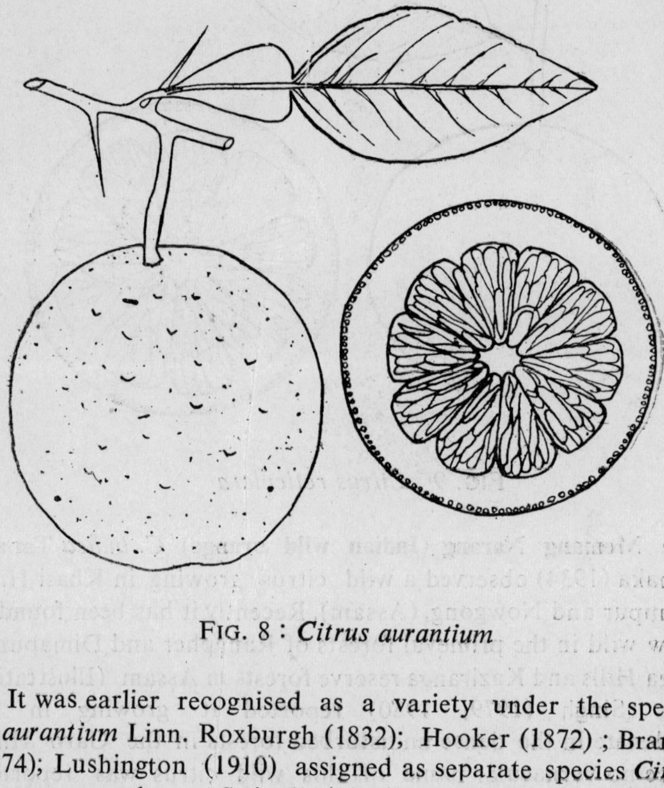

FIG. 8 : *Citrus aurantium*

It was earlier recognised as a variety under the species *C. aurantium* Linn. Roxburgh (1832); Hooker (1872) ; Brandis (1874); Lushington (1910) assigned as separate species *Citrus Chrysocarpa* whereas Swingle (1919) recognized it as *C. nobilis* Lour. Subsequently *C. reticulata* for mandarin oranges has been accepted (Illustration C1).

The tree is small spiny with dense top of slender branches ; Leaves : ovate, pointed; Petiole : slightly margined, articulated, deep green; Flower: white, solitary, small to medium, smaller than those of *C. sinensis*; Fruit: medium size, depressed or flattened globose, orange yellow to deep orange; Segments : 8-13 usually, 10 separating readily from one another; rind, thin, loosely attached, easily removed, distinct aroma; Seeds: smooth beaked, cotyledons greenish. It blooms annually and generally considered to be hardier than the sweet orange (Fig. 9).

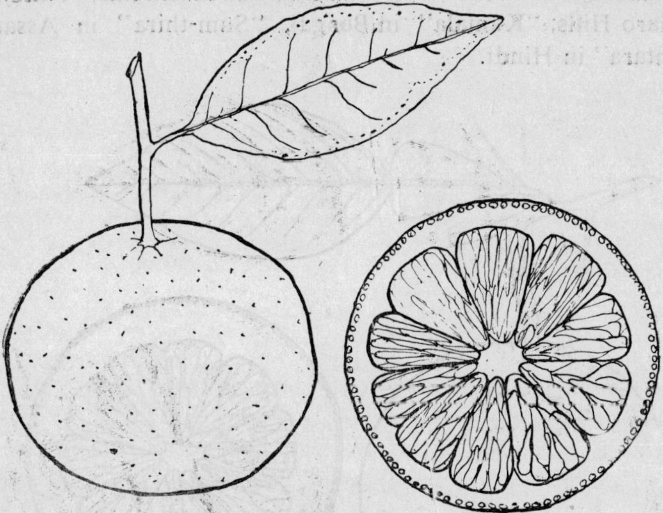

FIG. 9 : *Citrus reticulata*

The Memang Narang (Indian wild orange) *C. indica* Tanaka Tanaka (1934) observed a wild citrus growing in Khasi Hills, Manipur and Nowgong, (Assam). Recently it has been found to grow wild in the primeval forests of Rangpher and Dimapur in Naga Hills and Kaziranga reserve forests in Assam (Illustration C2). Singh (1979, 1980) reported it growing in the wild state in the dense undisturbed forests in the Garo Hills. Since no vernacular name of this wild citrus was reported, this was called the Indian wild orange. The recent intensive survey of this region undertaken by me revealed that this citrus growing wild in the Garo Hills is commonly called as 'Memang Narang' (Memang=ghost, Narang=orange in Garo language). I therefore consider that it should be termed henceforth as 'Memang Narang' (Illustration B4).

Distinct diversified types of 'Memang Narang' have been observed growing in a wild state in the undisturbed forests of Tura range of Garo Hills. These ranged from thorny densely foliaged bushes to tender climbers. The extent of variability encountered in these forests in the wild state unequivocally suggests this part of the world to be the native place of Memang Narang. It is characterised by small spiny bushes with height ranging from 3-6 feet, spreading branches arising almost right angles, glabrous, 2-4 cm long pointed thorns, moderately dense foliage; Leaves: oblong, gradual tapering at both ends, apex acute, medium size, upper surface somewhat deep green than the lower surface, crenate margin, reticulated venation; venation relatively more prominent and distinct from other citrus types; size: 6.0-11.5 cm long; width : 2.0-4.5 cm; Flower: bisexual, solitary rarely in pairs, pedencle short 3-4 cm long; Sepals : greenish white 4-5 lobed glaborous cupular; Petals : 5, imbricate, pure white 9-16 mm long, 3-7 mm broad, apex somewhat obtuse, oil glands present; Stamens : many filaments, white joined from base forming a tube, top free, 8-12 mm long, anther yellow, 2.0-3.5 mm long; Ovary : almost 4.0 mm long and 3.0 mm broad fixed on a disc, 8-11, locules; Style : white 8-10 mm long, stigma capitate; Fruit: depressed, globose to spherical size, small 23-35 mm height, 25-38 mm diameter, colour deep orange, surface smooth, oil glands present, very thin rind, usually soft, adhere, slight oil glands prominent, aroma distinct, mesocarp thin, white spongy, soft; segments 8-11, adherence between carples slight, juice scanty, practically inedible, taste very sour; Seeds: prominent and many, cotyledons green. The flowering season is sometimes from September to February. The plants are relatively hardier, obviously they show resistance to diseases and pests. The plants were relatively more prolific in the dense forests (Fig. 10).

The Pommelos, *C. grandis* Linn Osbeck (2n=18)

The pommelo is not of Indian origin. It is considered to be indigenous to Polynesia and Malay (Hooker 1872 and Webber 1946). It was introduced to India from Java. No mention has been made about it in Memoirs of Babar. Brandis (1874) reported that it was growing in the Garo Hills. Pommelo has been called by various botanical names. Roxburg (1832) and Lushington (1910) referred to pommelo as *C. decumana* whereas Chatterjee (1948) indicates that it is more proper to refer it as *C. maxima* (Burm). Recently, Swingle (1946) and Tanaka

(1937a) considered it best to retain the oldest valid name *C. grandis* (Linn; Osbeck) for this fruit.

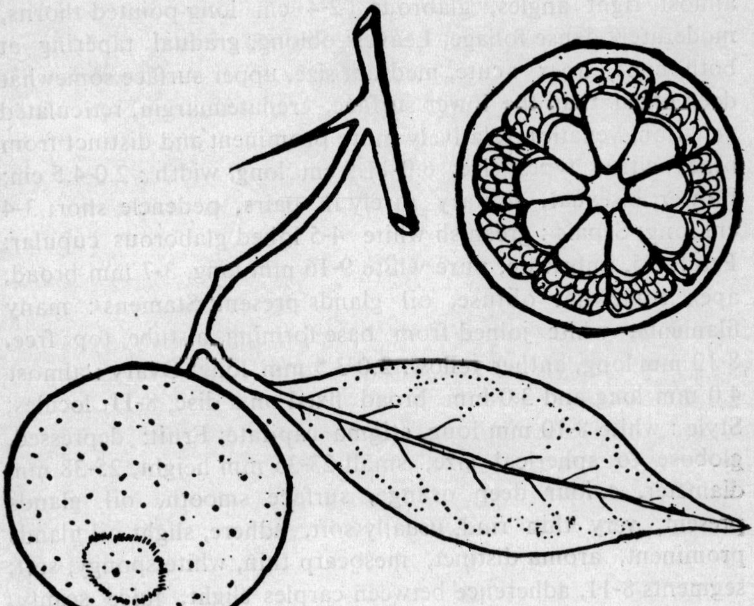

FIG. 10 : *Citrus indica*

The common name used as 'Soh myngor', 'Soh'-saw', (Khasi Hills), 'Jambura' or 'Matu' (Cachar); 'rabaa-tenga' (Assam); 'Batabi' (Bengali); Chakotra'; 'maha-nibu; sadaphal'; (Hindi); 'obokotru' (Gujrati), 'Pumparamasam' (Malyalam), 'Chakotre', 'Sakkota' (Kannada) and 'Pambalimasu' (Tamil).

The trees are medium tall with compound, rounded or somewhat flattened head (Illustration B5); Leaves: large 10-20 cm long, ovate-oblong, slightly pubescent on the underside, apex and base rounded, broad winged petiole, margins entire; Flowers : large, axillary recemes; Petals: white 4-5 thick dotted with large oil glands; Stamens: many; Fruit: very large lemon to orange colour red; rind : very thick, white, spongy, smooth prominent oil glands; segments : large, usually 10—14, juice vesicles large, loosely adhering, pulp light coloured or pink, tough, bitter or acidic; Seeds: many, large, flattened, monoembryonic (Illustration B6). These are generally heavy yielders. These grow on varied elevations to an altitude of 1600 metres (Fig. 11).

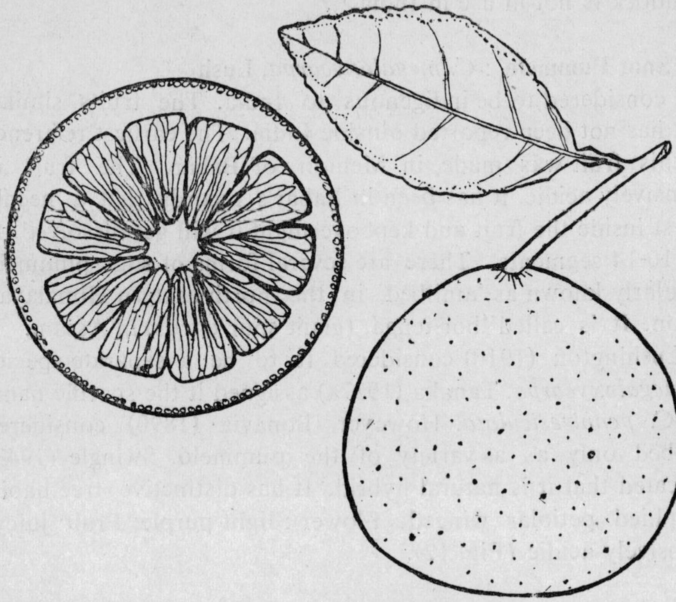

FIG. 11 : *Citrus grandis*

Grape fruit : *C. paradisi* Macf. (2n=18)
The grape fruit closely resembles pummelo and is recognised by some as *C. grandis* var. uvacarpa. It was recognised later as separate species as *C. paradisi* Macf. [Swingle (1946) recognizes it as a 'setellite' species. This may be a hybrid between pummelo and sweet orange. The grape fruits have been introduced to Asia. This must have occurred in the West Indies sometime before 1750. It is widely cultivated in the tropics and sub-tropics. The grape fruit tree and pummelo are distinguished by means of a group of characters rather than one factor alone. The leaves are relatively smaller, slightly lighter in colour and less leathery, little pubescence, smaller winged petiole. The

fruit is usually smaller, thinner, rind, more juicy, somewhat different flavour.

American have contributed to the confusion of names as they adapted the term 'pomelo' for the grape fruit, pronouncing the word similar to pummelo. Thus they distinguished pummelo (Shaddock) from the pumelo (grape fruit). The term Shaddock is not in use in India.

The Sour Pummelo : *C. megaloxycarpa*, Lush

It is considered to be indigenous to India. The fruits similar to it has not been reported outside India. The earliest reference to this fruit was made in Memoirs of Babar. The fruit is intensively acidic, it has been in Babar's records that a needle thrust inside the fruit and kept overnight it will get dissolved. It has 10-14 segments. There are several forms of sour pummelo popularly known as 'amilbed' in the north-eastern Himalayan region. It is called 'bor-tenga, (giant sour citrus) in Assam.

Lushington (1910) considered it to be a separate species *C. megaloxycarpa*. Tanaka (1937a) assigned it the specific name as *C. pennivesiculata*. However, Bonavia (1890) considered amilbed only as a variety of the pummelo. Swingle (1943) indicated that it is natural hybrid. It has distinctive tree habit, atrophied petiolar winged; Flower : light purple; Fruit: juicy, intensively acidic (Fig. 12).

Papda

There are three recognizable species which could be placed under papda, these are *C. macroptera*, *C. ichangensis* and *C. latipes*.

C. Macroptera Mont.

They do not have pleasant edible fruits. The fruits have pulp vesicles which contain droplets of acrid oil. They have distinct characteristics of leaves, flowers and fruits.

The regional common names for the *C. macroptera* are 'Sohkwit' (Khasi Hills), 'Satkara', 'Sat' presumably refers to multiple of seven as the fruit contains fourteen carpels and 'Kara' part (Assam, Mikir) and 'Chamarbaphang' (Cachar Hills). It is considered to be a native of South-East Asia and is commonly found in Thailand, Indo-China, Philippines, New

Guine and Polynesia. This species was discovered by Father Montrouzier (1860); Bailey (1903) named it *Citrus papuane*. Tanaka (1937a) placed it as *C. macroptera* var annamensis. However Swingle (1946) placed it as *C. macroptera* Kerii. Kanjilal *et al.* (1936) identified satkara of Assam as *C. hystrix*, D.C. to which Bhattacharya and Dutta (1956) did not agree. They claimed that no real *C. hystrix* existed in Assam. The fruit identified by Hooker (1872), *C. hystrix* D.C. as was doubtful. Bhattacharya *et al.* (1956) considered it to be actually *C. latipes*.

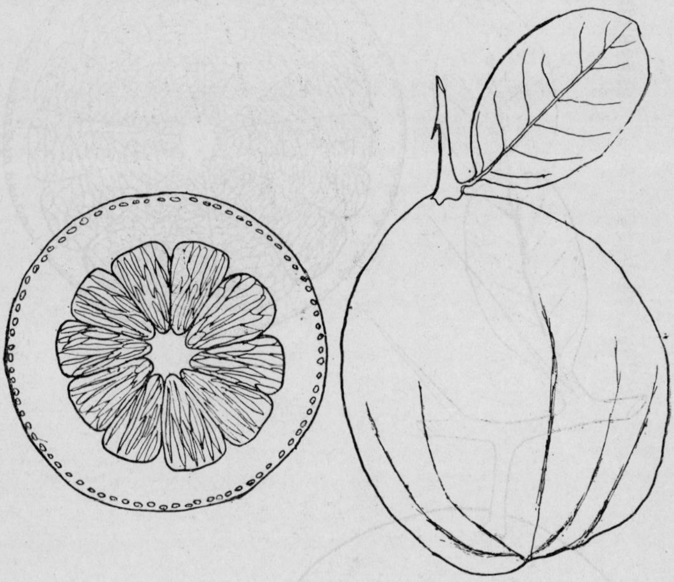

FIG. 12 : *Citrus megaloxycarpa*

It is a very tall tree 23-50 feet high, densely foliaged, thorny, prominent, obovate winged petiole (Illustration B7). The petiole is commonly larger than the blade. The blade is ovate elliptical shaped, the leaf including petiolar wing 8-17 cm long; Flower : small, generally clustered, petioled white; filaments, free, pistils stud shaped, style very short; Fruit : round oblate surface, smooth, pitted pulp vesicles small, contain oil droplets; 13-15 segments, usually 14. Seeds : thin. It is found abundantly growing in the Cherrapunji and Shella area of Khasi Hills, Mikir and North Cachar Hills in Assam (Illustration B8). These were found well distributed between Cherrapunji and Shella areas. These were found growing in wild state, on the road side also. The detailed description of the specimens during the present investigations was found to be similar to that described by Bhattacharya and Dutta (1956) (Fig. 13).

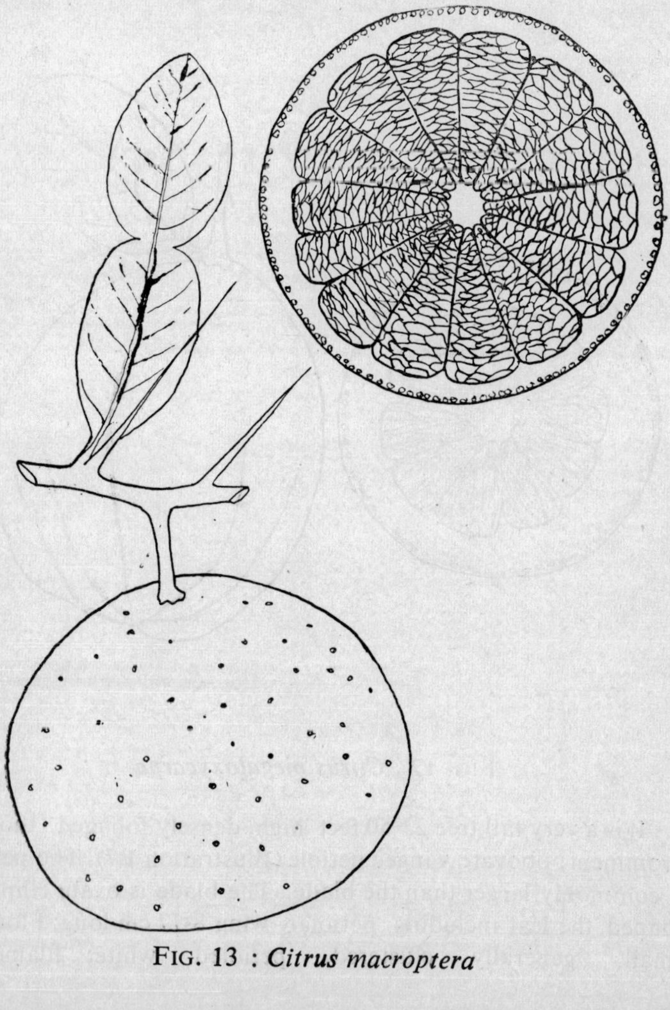

FIG. 13 : *Citrus macroptera*

C. ichangensis Swingle 2n=18

It is considered to be a native of south-western China, probably Ichang region in China. It is found growing in a wild state in the plateau of Shillong and other parts of the Khasi Hills and Naga Hills at an elevation of about 1800 metres. It is called 'Ketsa shupfa' by the tribe Angami of Nagaland.

It is a small-medium sized tree 15-20 feet long, angular branches, slender spines; leaf narrow with oblong broadly winged, dark green articulated, ovate-lanceolate, 5-6 cm long and 2-2.8 cm broad, candate apex, margin slightly crenate, petiole nearly as large as the blade, veins running parallel to one another particularly in petiole, evenly rounded apex, abruptly narrowed at the base, veins almost parallel in the petiolar wing; Flower : white, stamens many; Fruit : small, oblong, smooth, pulp vesicles small, sour and acrid taste ; segments 7-9; Seeds : large thick cunceate-ovate white. It is grown as an ornamental plant at some places. Frost resistant, it is useful rootstock and can be crossed with cultivated species (Fig. 14).

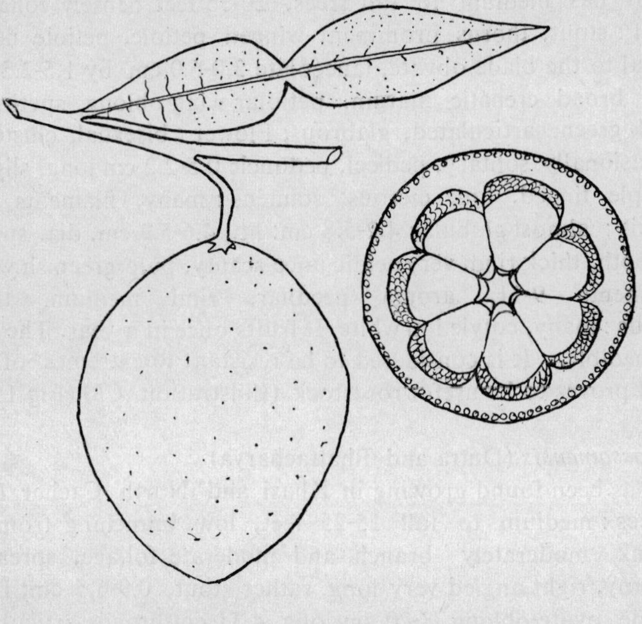

FIG. 14 : *Citrus ichangensis*

C. latipes (Swing) Tan.
The species has been found growing in a semi-wild state in the Shillong plateau, Mawprem, Mawphlang, Pynursla, Cherrapunji and Shella areas of Khasi Hills. This was also found in Naga Hills. This has been observed growing oocasionally in the courtyards in the houses and along the road side between Cherrapunji and Shella area. The detailed description of the specimens revealed that these were precisely similar to that described by Dutta and Bhattacharya (1956). It may be considered to endemic to Eastern Himalayan region. It has been also reported by Tanaka (1937) and Swingle (1946). It has been found growing to an altitude of about 1900 metres. It is commonly called as 'Soh-kymphor-shrieh' (Soh=fruit, Kymphor=shrieh-monkey) in Khasi language, a fruit of monkey (Illustration B9).

Swingle (1913) placed it under the species *C. ichangensis* sub-species latipes. Subsequently he changed it and classified it to a separate species, *C. latipes*. Tanaka (1937) also placed it under the species *C. latipes*.

It has medium to tall trees, 20-35 feet densely foliaged, Leaf : stout, thorns prominent winged petiole, petiole nearly equal to the blade obvate, lanceolate 3.0-6.0 cm by 1.5-2.5 cm and broad crenatic margin, petiolar wing oblong, spethulate dark green, articulated, glabrous; Flower : bisexual, clustered, occasionally solitary, pedicel, peduncle 0.5-2.2 cm long, slightly purple tinged, four merous; stamens : many, filaments free; Fruit : almost globular 4.2-5.8 cm. ht., 4.6-5.9 cm. dia. surface smooth, thick rind, very acidic juice scanty, pulp greenish white; segments 9-11, aroma peculiar, rind medium thick; Seeds : many, cotyledon white. It fruits once in a year. The fruit is inedible. It is considered to be resistant to extremes of cold and proves to be useful rootstock (Illustration C3) (Fig 15).

C. assamensis (Dutta and Bhattacharya)
It has been found growing in Khasi and North Cachar Hills. Trees : medium to tall 15-25 feet, low branching from the trunk moderately branch and moderate foliage, spreading thorny, right angled very long, rather stout, 0.9-6.5 cm; Leaf : blade ovate-oblong, 8-20 cm long, 4-11 cm broad, articulated, deep green, very glossy, leaf aroma, strong and distinct, oil

glands in each notch, apex obtuse, small ovate, sphatulate in shape, crenate margin, winged petiole; Flowers : medium to large, bisexual, glomerate 4-8 (never solitary), cymes, pedicle short; Sepals : cupular greenish white; Petals : 4-8 deep purple; Stamens : many, polyadelphous, some free; Ovary : globose, 10-15 locules; Fruit : medium sized, smooth yellow, spherical pulp fusiform, rind medium thick prominent, oil glands, aroma strong, 10-15 segments, adherence strong, pulp vesicles white, abundant juice, juice intense sour; Seeds : many, cotyledons light greenish. There is strong distinct aroma of leaf and rind of fruit resembling somewhat ginger smell. It flowers generally in between the latter part of September and November (Illustration. B10 and C4).

FIG. 15 : *Citrus latipes*

It is called-Soh-Sying' (Soh=fruit, sying=ginger) in Khasi Hills, Ada-Jamir' (Ada=ginger, jamir=citrus) in Cachar Hills. It has been suggested that it may be placed as a variety of *C. hystrix* D.C. as 'ada-jamir' resembles closely *C. hystrix*. It was pointed by Tanaka (1937c) that *C. hystrix* did not occur in India. Bhattacharya and Dutta (1956) found that 'ada-jamir' differs from *C. hystrix* D.C. distinctly and hence it was placed under a separate specific rank, *C. assamensis*. 'Ga janimma' is found in South India. It has a pleasant acid juice and is commonly used as root stock. Tanaka (1954) identified it as *C. pennivesiculata* and considered it the same as 'attarra' which Bonavia (1890) reported in Madhya Pradesh. It resembles closely 'ada-jamir' or 'Soh Sying' of Khasi Hills. Dutta and Bhattacharya (1951) reported that 'ada-jamir' is distinct from 'ga-janimma' and identified as *C. assamensis* (Fig 16).

FIG. 16 : *Citrus assamensis*

6

Sources of Genetic Wealth

CENTRE OF DIVERSITY

THE ERADICATION of natural habitats for utilitarian needs has drawn the immediate attention of mankind to collect plant diversity. There are several such problems but certainly no specific recommendations can be given to fit all cases. The problem of collection and conservation of plants did not attract the people in the earlier days due to the smaller number of human beings that peopled the earth. The present situation is alarming as enormous amount of destruction of plants is being done due to man's multifarious development activities. Presumably, the conservation of plants cannot be ignored any longer. These collections may be made in the form of seed, cutting, bulb or any other storage organ or in some cases, it may even necessitate the removal of the entire individual from its natural habitat.

The various sources of gene pools have been indicated in Fig. 17. It is essential to review the past collection activities and to know the precise sites of genetic diversity for the systematic collections. Alphonse de Candolle (1883) was the first scientist to study 247 species of cultivated plants and speculated their origin employing four kinds of evidences, history, philology or linguistics, botany with particular reference to geographic distribution of wild relations and archaeology. He stressed that for a satisfactory answer, evidences from all possible sources should be employed. His conclusions that cultivated plants originated in a few places were based on Swiss lake dwellings, a region inhabited by man thousands of years ago, which show cultivated types of plants, while

Australia which was invaded by man recently lacks such varied forms. He was possibly the first to indicate regions where plant domestication might have taken place. These are China, South West Asia and Egypt and Tropical Asia. The plants which adapt better in the man-made environment are called domesticants. He suggested that abundance of a species in a region does not necessarily indicate the region to be its centre of origin.

Nicolai Ivanovic Vavilov (1927) developed a theory of the centre of origin or gene centres. He encountered rich diversity in certain regions of the world, while collecting cultivated plants and their wild allies. He concluded that the primary centre is noted by the high frequency of dominant alleles, the frequency of recessive alleles is increased towards the periphery and obviously the diversity is decreased, thus these develop secondary gene centres. The regions may overlap for cultivated plants. Vavilov (1931) distinguished seven gene centres, these are :

1. China
2. India
3. South-West Asia
4. Mediterranean Sea coastal and adjacent regions
5. Ethiopia
6. Central America and South Mexico
7. South America.

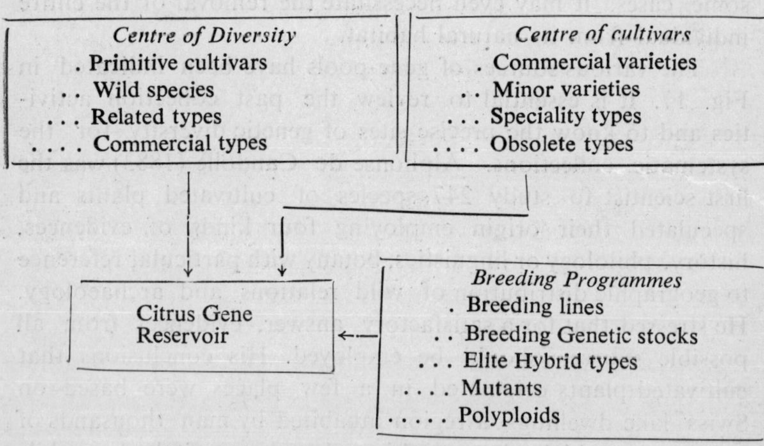

FIG. 17: *Sources of new-gene pools*

These centres lie in the mountains between 20° and 45° in mostly temperate climate areas. However these centres are separated by vast deserts. It was felt that these regions were developed independently due to differences in agricultural methods and domestic animals. Vavilov developed the concept of gene centres which was modified at later stages. He identified 19 agro-ecological groups in the Old World barring North and South Africa and tropical Asia. He suggested that the plant types within a group have some characteristics in common. These characteristics can conveniently be used as guidelines for the distribution of particular species.

The number of cradles of agriculture has been discussed in depth in the past by various workers. Vavilov suggested many, whereas Saucer (1952) indicated two, one for the Old World : Burma and adjacent areas; and one for the New World : Central America. Darlington (1952) also pointed out two, one for the Fertile Crescent of the Near East and the other for Mexico. The nuclear areas are speculated to be the source for spreading agriculture over the Old World and New World. Possibly, the new centres of plant domestication were developed after the introduction of agriculture. This led Darlington and Janaki Ammal (1945) to suggest 12 centres of origin.

Anderson (1960) divided the agricultural into floral and non-floral seed crop agriculture in Central Africa and a pole of floral agriculture in Indonesia. He speculated that floral agriculture spread into Oceania, Japan, China, India and Afghanistan. Whereas, the non-floral continued to remain in Africa. Curiously enough, the Africans do not seem to be interested in flower and ornamental plants. On the contrary, he indicated that even the poor man shows interest by way of growing some ornamentals in the region of the floral type. Subsequently, the claims of the African cradle were ruled out by various workers.

Zhukovsky (1968), pointed out that a large number of crops originated outside the centres of origin suggested by Vavilov. He put forth the idea of 'megagene centres' which are 12 scattered throughout the Old World and the New World. Recently, Harlan (1971) distinguished two types of origins of centres—one 'centre' and the other 'non-centre', in the non-centre, agriculture was introduced after which many indigenous crop plants were domesticated. The non-centres involve a larger area. He classified these two as under :

Centre	*Non-centre*
A1—Near Eastern	A2—African
B1—Chinese	B2—South Eastern Asian and South Pacific
C1—Central American	C2—South American

Various types of classifications for centres of origin have been suggested but the classification suggested by Vavilov still forms the basis of the subsequent work. The concept of Vavilov can be modified on the basis of more extensive information available on genetic diversity of crop plants.

In one way or the other cultivated plants must have arisen from wild types but this has happened at very different points in time. Many cultivated plants are very old, others are of recent origin and still others remain mainly at the wild stage. There are notable cases to demonstrate that plants may return from the cultivated stage to a wild species that were of importance earlier; for instance, certain dye and spice plants may now have a troubled existence. Other cultivated plants are still in culture but one threatened to be superseded by new types, Domestication brings manifold changes in the plant characters, the possible changes are enumerated as under :

1. Preference of new agro-climatic conditions.
2. May lose the dispersal mechanism, partially or completely.
3. Shattering of seeds is reduced.
4. May change flowering period, generally towards earlier flowering.
5. May have changed to annual from perennial.
6. The period of seed dormancy may be reduced or eliminated.
7. May have changed the flower morphology, change from self-incompatibility to self-compatibility.
8. Change in tendency from partial or complete cross-fertilization to self fertilization.
9. May be multiplied vegetatively.

B. 8 : Citrus macroptera

B. 9 : Citrus latipes

C. 1 : Citrus reticulata

C. 2 : Citrus indica

10. May have changed to increased size of flower and seeds.
11. May have changed to seedless parthenocarpic fruit.
12. May have lost photo-period controls.
13. May have lost completely or reduced considerably the protective coverings and defensive adaptation such as spines, thorns and hairs.
14. May have improved in quality like protein content and palatability.
15. May have reduced in sturdiness and increased in susceptibility to diseases and pests.

Zeven and Zhukovsky (1975) stated that the citrus fruits have an origin in the Indo-Chinese-Indonesian centre, Chinese-Japanese centre, Mediterranean centre, Hindustani centre and also some are found in the African centre, South American centre, Australian centre and Central American and Mexican centre. The north-eastern Himalayan region is considered the natural home of the citrus fruits. Several primitive cultivars, wild types, related types and commercial types have been encountered in this region.

SOURCES OF NEW GENE POOLS THROUGH CENTRE OF CULTIVATION AND BREEDING PROGRAMMES

Nomadic man strived hard to find out the various wild types of plants to fulfil his requirements. Domestication of plants made it possible for man to settle down and thus relieved him from the compulsion of a nomadic existence. It is speculated that the nomads may have started hunting first in the woodlands and near water for fishing. This made them identify the sites which subsequently they frequently visited. This might have led them to seek occupation for a limited period. Gradually when these resources at these sites were exhausted they moved to other areas. These people might have brought some plant parts around their dwellings, some of these findings suitable environment started growing. By and large, as man moved to disturbed areas, they also carried with them some useful plants. The refuse heaps could be another useful site where plant parts had the chance to grow on such fertile land. The present day plants are thus developed from their wild forms. In several cases the cultivated forms can easily be traced back to certain wild species.

The cultivation of the plants facilitated in laying down the foundation of the present day civilization. Human civilization is absolutely dependent upon plants for obtaining cereals, vegetables, fruits and other materials required for its very existence. These requirements made man as the hunter of plants. The process has gone slowly for several millennia, but during the recent past people have taken keen interest.

Commercial varieties of several crop species, i.e. citrus fruits, including their varieties of minor importance, speciality and obsolete types are found growing at one or the other parts in the centre of cultivation. The commercial varieties are well adapted to the local conditions. These are generally improved types through various selection and breeding procedures utilizing the natural variability. These improved types are combinations of several desired characteristics both for quantity and quality. Some of these are improved types for specific quality purposes or adaptable to certain regions.

The plants when moved from their natural habitat to new environments, showed that some could adapt themselves while others could not. The adaptation to new environment led plants to undergo certain morphological changes in the plants and their reproductive organs. This resulted in more variants from the wild types. By and large, as time passed man had to learn to grow the plants for his requirements. He started selecting a few plants to the exclusion of others. Subsequently, he preferred some characteristics of plants and obviously neglected others. These semi-directed processes led to the improvement of the plant species. As man developed, his demand also grew with him. He started to learn the art of improving the crop more systematically. He gradually learnt to create diversity through his means by way of hybridization, polyploidy and mutations. The available variability could be thus carefully collected and conserved.

Methods of collections of citrus fruits

The operations pertaining to systematic collection programmes are enumerated as follows :

1. Communication with the officials of local departments.

2. Steps in collecting materials.
3. Handling of collections.
4. Cataloguing and storage of collections.

1. *Communication with official of local departments*

For a systematic approach to the subject, it is desirable to involve the official of the local departments particularly of horticulture, forest and agriculture. Their involvement is absolutely essential in understanding the areas, particularly the inaccessible pockets for collecting the desired germplasm. The local contact, like the Horticultural Officer, District Forest Officer, District Agricultural Officer, other Extension workers and village chiefs or nokma, for their necessary assistance at the time of surveys and collection of citrus fruits cannot be overrated. The local authorities can provide useful information regarding the distribution and even slight changes in the extent of variation. This information is extremely useful for the preparation of travel plans to such areas. This may be supplemented with proper maps indicating the route of travel.

Citrus fruits are grown in tropical and subtropical regions at varied elevations throughout the world. In India most of the genetic variability is confined to the north-eastern Himalayan region. In the case of citrus fruits, the natural variability prevails in the forests, thus making the job all the more difficult. It is expected that the explorer be well acquainted with the local cultivars, commercial types, wild types, primitive types and related species. It is particularly of interest in the case of citrus fruits having a number of species each with enormous types. Some of the *Citrus* species are still faced with controversy with regard to their systematic botany. These fruit crops are also faced with peculiar problems of being propagated through clones coupled with large size and longer life.

2. *Steps in collecting materials*

It is imperative to review the past activities carefully including their evaluation and method of conservation. Its usefulness lies in planning the field collections as follows :

1. In the areas where past efforts provided useful

materials scientists generally are tempted to gather the working collections, and thus their collections consist of commercial types. But it is the primitive and particularly the wild types that are reported to be of special value and are of utmost importance and need particular attention today.

2. Special efforts should be made in the remote areas where negligible collections have been made, and where seemingly the primitive types may be still found in an uncontaminated and pristine form.
3. The unique areas where a few citrus types tolerate adverse factor(s), water, pest, disease, soil and sunlight etc. may be surveyed.
4. The inaccessible remote areas of the forests which seem to possess great potential for primitive and wild citrus types need our undivided and immediate attention.
5. The unexplored areas for the related types or species also require special efforts.
6. The following information and procedures may be adapted for obtaining the rough estimates regarding the distribution pattern of a species :
   - (a) The examination of the records of the various research stations dealing with citrus fruits will be a useful exercise. While doing so, one should not neglect to study thoroughly important herbaria for citrus fruits and their related types.
   - (b) The available regional statistics regarding production may be utilized for providing the rough estimates of the distribution of species.
   - (c) Possibly aerial surveying techniques supplemented with ground reconnaissance can be utilized. These techniques can also be used for providing information regarding the characteristics like flowering and fruit maturity which are essentially required for planning the expedition.

In the case of citrus fruits, scientists are not in agreement with the precise specific positions of some types, thus it has resulted in different names being given for the same type or vice versa. The nomenclature of these types could be verified and similar types be brought together while compiling such types of information.

It is worthwhile to make a trip to the central places before such expeditions are undertaken in order to acquaint oneself with the local problems. It is required that these teams should be properly equipped. They may possibly need the following articles :

1. Jeep with the necessary accessories like trailer, etc.
2. Camping equipment like tents which should provide protection from heavy rain etc. There should also be provision for sufficient searchlights which are useful to scare wild fauna.
3. Sufficient food supplies, cooking utensils, stove etc.
4. First-aid box.
5. Pruning knife, grafting wax, grafting iron, shears and a small mallet.
6. An instrument box containing scissors, magnifying lens, etc.
7. Altimeter and such other requisite equipments.
8. Record sheets, cloth bags, envelopes, labels, ink for printing and writing, punching and stapling machines, balls of thread, writing pads and postal stamps etc.
9. Insecticides and insect repellents.

In case it is intended to cover relatively larger areas which are inaccessible such as the north-eastern Himalayan region, it will be desirable to operate more than one team depending upon the facilities in hand. It is desirable to vist these areas at least once when these types are in the flowering stage and a second time when these bear fully matured fruit. It may thus be essential to formulate the programmes accordingly. Many of the *Citrus* species have a common range of flowering period though it may differ in some cases. In addition the forest areas are accessible only for a limited period in a year. Firstly these are not accessible during the rains; and secondly, due to the luxuriant growth immediately after the rains.

SAMPLING

The varietal diversity will determine the number of collections to be made. The best estimate of the number of samples can be made at the time of travel while studying the existing variability. Special attention should be focussed on the remote areas to collect primitive varieties and related wild species of citrus fruits. The object of the collector should be to include as much variation as possible in a sample. The explorer should take equal interest in the primary and secondary centres of diversity. It is reported that the periphery of their distribution should in no case be neglected as in such areas varied types with important traits like disease and pest resistance may be found. These genotypes are adapted to conditions not found in the central area of distribution. Although in the crops as highly cross-pollinated as fruit crops, it is desirable to collect random samples, due to the practical limitations, it is advisable to resort to the systems where random and biased sampling are combined. It is suggested that in the first instance the grid sampling covering the larger area should be undertaken. This will indicate the range of variability in the area. This is desired as in some cases the individual fruit types are widely dispersed. If vegetative parts are collected, it becomes bulky and rather difficult to maintain variability. This technique also requires the proper rootstock for their maintenance. It may be essential to resort to biased sampling in some cases where the explorer is interested in some particular traits. The random samples should be collected during the intensive surveys. The citrus types which are under cultivation are generally propagated vegetatively and hence are homogeneous. In such cases one clone for each collection may be desired. It should be remembered that several commercial citrus fruit types have arisen through mutation or bud sport. It is therefore suggested that for the proper sampling, the collections should be made to cover the entire area. Seeds are normally more convenient to collect and are preferred. The vegetative materials for collections are preferred particularly when specific parental genetype is required. Whenever the seeds are collected, it is advisable to record the surrounding trees as potential pollen parents. Sykos (1975) has stressed that these collections should be identified for their proper taxonomic position by growing them under uniform conditions and thus evaluating.

3. Handling of collections

It is essential to maintain the detailed records of these collections. Notes should be recorded on the locality, local name, botanical name, history of the variety, rainfall, temperature, altitude and other relevant information in addition to maturity, disease and pest resistance and other useful characteristics of fruit and tree. The records pertaining to collections should also include information such as collection number, date of cultivation, owner's name, situation of site of collection, soil including its topographical features, cultural practices and seed, seedling or vegetative propagation source and other relevant significant information of future value. The distinguished characteristics of the type regarding tree such as thorns and leaf including its shape, the petiole, fragrance etc. and the fruit characteristics such as size, segments, number, peel and rind characteristics, pulp fragrance etc. may be recorded.

4. *Cataloguing and storage of collections*

Some of the characteristics may be recorded at the site while others can be recorded in the laboratory. Each sample should be photographed. The samples may be in the form of seed (fruit), or vegetative parts. If bud sticks and graft wood are collected, these should be properly packed in plastic covers and kept moist. These are relatively less bulky and are convenient for transportation. These can be utilized by budding or grafting. The appropriate season for grafting is when the trees are dormant. Budding can be more profitably done at the time when the sap is active in the rootstocks. The desired rootstocks have to be arranged in advance. A sample of a field record sheet is shown in Table 7.

It may not be possible to record all the desired characteristics in detail at the site due to the paucity of time. It is convenient to score the characteristics particularly those to be recorded at the time of sampling. For instance, in the case of plant characteristics, tree: tall=1, medium=2, short=3; thorns may be scored, long pointed=1, medium=2, small=3; leaf: large=1, medium=2; small=3; petiole: large=1, medium=2, small=3 absent=4; fruit characters such as large=1, medium=2, small=3; taste=sweet=1, medium=2, sour=3.

Similarly abbreviations and score for recording the other characteristics may be conveniently used in the laboratory. The proper record for the abbreviations and score used in recording the data should be kept.

TABLE 7: **Citrus fruit—field record sheet**

Institution ————————————

Collection No.———————————Accession No.————

Latin Name——————————

Vernacular name——————————

Locality: State———————————District————

Village———————————

Producer's name——————

Altitude——————————Longitude————

Rainfall ——————Av. Temperature————

Soil————————————

Cultural practices ——————————

——————————

History——————————

——————————

Material: Seed ——————Vegetative parts———Plant———

Status: Cultivated Semi-wild Wild

Frequency Abundant Frequent Occasional Rare

Habitat :

Plant characters: Tree Tall Medium Short

Leaf Characters: Shape

Winged: Prominent Medium Absent

Leaf blade/wing size (approx).

Fruit characters:

Skin Loose Tight

Rind

Segments

Pulp

Taste Sweet: Sour: Acidic: Semi-acidic

Uses ——————————————

——————————————

Disease Symptoms————————————————
————————————————
————————————————

Remarks————————————————————
————————————————————
————————————————————

Date of Collection :

Photograph No.

Herbarium sheet No.

Collector's name.

7

Maintenance and Propagation of Citrus Fruits

CITRUS fruits can be propagated by seeds, budding, grafting, cutting or layering. The layering is not much in practice whereas the budding, by far, is the most common method.

PROPAGATION BY SEEDS

Seedling trees do not produce fruit true to the type from which the seed is obtained. Some of the prominent commercial varieties have been selected from seedling trees. It is expected that some of them may not be prolific and producing desirable fruit. The preference for this method is gradually dwindling. The seed should be obtained from healthy trees with desirable characteristics. Poorly developed seed should not be used. Seeds from the fruit may be obtained by cutting the fruit and squeezing them into a sieve. The seeds may also be washed free from the pulp afterwards. The seeds may also be obtained by allowing the fruits to rot. Running water is used to wash the seeds free from the pulp and then sieved. Poorly developed seed should be allowed to float away and be discarded. The cotyledons separate and lose viability in case the seeds are too dry. In case these are planted immediately, these may be stratified for a short time in damp sand. When the seeds are stored for some considerable time, these should be dried quickly after washing and mixed with an equal bulk of finely ground charcoal and kept in air-tight tin cans.

PROPAGATION BY CUTTINGS

The cuttings of the citrus fruits like orange and citron are difficult to root and have little practical value. Matured wood of one or two seasons should be used for cuttings. These cuttings may be about 5 inches long with 3 or 4 matured buds. These should have clean smooth cut, this facilitates the formation of callus, which must take place before roots are developed. These cuttings may be placed in clean washed sand in rows 2-3 inches apart and about 2 inches apart in the rows. The cuttings may be transplanted to pots at the stage when these have developed roots and sent out shoots bearing 2-3 leaves. Subsequently they may be allowed to develop into trees or may be grafted or budded.

PROPAGATION BY BUDDING

The budded or grafted trees proved to be adaptable to new conditions, divergence and specialization. These trees bear fruits earlier. The varieties maturing at different seasons can be planted which are more profitable. These also produce desirable quality and uniform sized fruit with higher yields. The root stocks thriving well at varied soil conditions can be fruitfully utilized. These rootstocks may be useful to avoid certain diseases caused by fungi and viruses.

It has been observed that some buds or grafts do not produce true to type and have resulted in fascinating notable bud variations. A large number of variations in shape and size of the fruit, peel colour and thickness and yield are observed in the single branch, a part or in the whole tree. These variations should not be attributed to bud sports alone. These could be as well due to the differences in root stocks and environmental conditions like soil, plant nutrients and moisture etc.

Budding is commonly practiced. The buds may be inserted easily in rootstocks any time during the growing period when there is sufficient sap. Generally the stocks are budded when two or three years old. When they have attained sufficient size the bark may easily be split. The point of union between the scion and rootstock is more susceptible to cold. If the point of union is well down towards the crown roots, the earth may provide useful protection or may be banked easily against the point of union during the cold. While doing so, every care should be taken that the point of union is not too close to the earth else this will increase the chances of diseases like foot root. The T-budding or inverted T insertion is a common method of budding.

The stock should not be only adaptable to the environment but it should be effective to the scion. The common effects of the stock to the scion are the rate of growth, ultimate size, hardiness, maturity period, vigour and quality of the fruit.

8

Conservation of Genetic Resources including *in situ* (on site) Preservation of Genetic Variability

EXTENSIVE studies have made it clear that the bulk of variability in cultivated plants is restricted to certain regions of the world and most of the regions are confined to the developing countries. It is all the more ironical that in regions of much remarkable natural richness, its people are the poorest in the world. The developing countries possess the enormous genetic variability in economic plants; it is therefore their responsibility to ensure that this useful variability is not lost due to the introduction of more productive varieties and land use patterns.

The need for conservation and its strategies is paramount. Conservation is defined by Budowski (1975), as "the management of resources of the environment—air, water, soil and living species including man—so as to achieve the highest sustainable quality human life; management in this context includes surveys, research, legislation, administration, preservation, utilisation and implies education and training". Early man was considered as an integral part of nature as his impact on the other components of the biosphere was indistinguishable. The social and cultural evolution of man has been marked by a growing conflict between him and nature. In other words conservation means the management of the human use of the biosphere in such a way that it yields the greatest sustainable benefit to the present generation at the same time aims to maintain its potential to meet the aspirations and needs of future generations. Precisely, conservation implies preservation, maintenance, appropriate utilization, restoration and enhancement of the natural environment. The living resource conservation is concerned with plants and animals and with inorganic matters on which they depend.

The living resources are renewable if they are conserved and are destructible if they are not conserved. Conservation should be treated as a way of life. Conservation and development are mutually dependent. The process of development should be so designed as to conserve living resources. Conservation will be more effective if it also meets short-term economic needs. There are about 1500 million people in the world who are dependent on wood, dung or crop waste for their fuel. Rural communities are primarily dependent directly on the living resources. These are the poor people and for them conservation is meaningless if it does not meet their immediate needs. Dung and crop wastes to the extent of 400 million tonnes are burned annually in different parts of the world. These could have been highly useful for the regeneration of the soils. There are many such activities which are intimately associated with the very existence of mankind and are liable to bring changes in the ecosystem. Development can be effective and more fruitful if it is guided by ecological, social, cultural and ethical considerations. It is thus inevitable to ensure sustainable utilization of species and ecosystems and to preserve the range of genetic diversity of plants and animals. The maintenance of ecosystems essential for food production and the health of human beings is vital for development of both developed and developing countries. The main ecosystems are also called the life supporting systems. The coastal and freshwater, agricultural and forest systems are the threatened life supporting systems.

FORESTS

The forests are known for the generous supply of food, fuel and fodder. The forests besides supplying clear water, control the regional climate. They protect the soil from severe erosion and floods. The watershed forests prevent the clogging of reservoirs, irrigation systems and canals. The watershed forests which are so valuable are being cleared for fuel, agriculture and communication purposes and are damaged by overgrazing. About 50% of the world's population, which live either in the mountain areas or on the adjacent plains are directly dependent on the management of the watershed ecosystem.

Land management and farming systems should be devised not only to produce food from the marginal agricultural lands but these should be capable of maintaining and improving the quality of the producing environment. These systems are termed as agro-forestry. Agro-forestry has been defined by Bene et al. (1977) and King and Chandler (1978) as "a sustainable land management system which increases the yield of the land, combines the production of the crops (including tree crops) and forest plants and/or animals simultaneously or sequentially on the same unit of land, and applies management practices that are compatible with the cultural practices of the local population". Thus these systems are devised to improve the site and at the same time optimise the combined production of forest crop and agricultural crop. Agro-forestry constitutes the following four systems (King 1979) :

1. Agri-silviculture systems

The system that permits land use simultaneously for the production of agricultural and forest crops.

2. Sylvopastoral systems

The system provides both the production of wood and rearing of domesticated animals. The animals are kept and grazed within the forests unlike agrisilvicultural systems where forage is grown in mixture with the forest trees which is harvested for feeding elsewhere.

3. Agro-sylvopastoral systems

This system is a combination of the former two systems, agri-silviculture and sylvopastoral system. This system permits the simultaneous production of agricultural and forest crops and rearing of domesticated animals.

4. Multipurpose forest tree production systems

This system provides the regeneration and management of

forest trees not only for the production of wood but also for the utilization of leaves and fruit for food and fodder.

The tree species selected to be grown in conjunction with agricultural crops should have *inter alia* the characteristics as follows :

TABLE 8 : **Comparison of India's land classification with that of some of the important countries of the world**

Country	Year	Total area Thousand Hectares	Percentage of total area: Agriculture Area: Arable land and land under permanent crops	Percentage of total area: Agriculture Area: Permanent Meadows & Pastures	Percentage of total area: Forest & Wood lands	Percentage of total area: Other lands
1	2	3	4	5	6	7
India	1969-70	306,042*	49.4	4.2	21.6	24.8
Belgium	1970	3,051	27.9	24.5	19.7	27.9
France	1970	54,703	34.9	25.5	25.6	14.0
Italy	1970	30,123	49.6	17.4	20.4	12.6
United Kingdom	1969	24,401	29.8	49.6	7.7	12.9
U.S.S.R.	1970	2240,220	10.4	16.7	40.6	32.3
Canada	1971	997,614	4.4	2.5	44.4	48.7
Mexico	1970	202,206	13.6	34.5	9.1	42.8
United States	1969	936,312	20.4	26.1	31.2	22.3
Pakistan	1969	80,394	23.9	6.2	2.3	67.6
Egypt	1970	100,145	2.8	—	—	97.2
Australia	1969	768,681	5.8	58.6	4.5	31.1
Denmark	1970	4,304	62.1	6.9	11.0	20.0
Argentina	1968	277,689	9.4	52.2	22.6	15.8
New Zealand	1970	26,867	2.9	47.7	23.4	26.0

* Relates to reporting area.

— = Negligible or not available.

Note : Figures are taken from F.A.O. Production Year Books, 1971, 1972 and 1973. In case of India, figures are based on Land Use Statistics for the year 1969-70.

Source : (Anonymous, 1977). Indian Agricultural Statistics, 1967-68 to 1969-70, Vol. II, Directorate of Economics and Statistics, Ministry of Agriculture and Irrigation, Govt. of India.

C. 3 : Citrus latipes

C. 4 : Citrus assamensis

B. 10 : Citrus assamensis

1. The root system of these tree species should be such that these should exploit the soil layers different to those used by the agricultural species.
2. The trees amenable to early wide spacement should be preferred.
3. Those should have self-pruning properties or at least they should be able to tolerate high incidence of pruning.
4. They should possess low index of crown diameter to bole diameter.
5. They should be of light branching habit.
6. They should have phyllotaxy which helps penetrate sufficient light to the ground.
7. The morphological changes in the above ground characteristics over time should be such that they retain or improve those features which reduce competition for sunlight, nutrient and water.
8. Their leaf flushing and leaf fall system should be advantageous to the growth of the agricultural crops in conjunction with which they are raised.
9. They should be capable of tolerating side shade, if indeed not of full overhead shade in the early stages of growth.

Developing countries with the highest population have to carefully plan their development and conservation strategies. Asia has the smallest area of 672 million hectares of 'potentially arable land' in comparison with 681 million hectares in South America and 734 million hectares in Africa (U.S. Rep. 1967). The comparison of India's land classification with that of some of the important countries of the world is indicated in Table 8. Denmark has the highest and Egypt has the lowest percentage of arable land and land under permanent crops. In Asia about 83% of potentially arable land is under cultivation. Obviously, there is little new arable land that may be utilized for food production. In order to keep pace with the growing demands of development it will be inevitable to consider the feasibility of increased production on land under utilization and from the land which is not considered arable. Agro-forestry systems can play a pivotal role in the socio-economic and physical sense in the areas like shift cultivation by the production of food and fuel and at the same time to conserve the ecosystem. In these areas forests were considered to act to restore soil fertility. More than 200 million people depend on shift ciltivation in the world. In India this practice is very common in the north-eastern Himalayan region. It is also prevalent in Orissa and Andhra Pradesh. It is reported by the National Commission on Agriculture in 1976, that 4.92 lakh tribal families involving 2.69 million hectares follow this system in India. However, at any one time the area under Jhuming has been reported as 453 thousand hectares. Although Sharma (1976) has indicated that the system was conceived in the Neolithic period around 7000 B.C., yet this primitive system of cultivation associated with numerous disadvantages continues to be followed in different parts of the world in the wet tropics (Schlippe, 1956; Conklin, 1957). The socio-economic, cultural factors, remoteness coupled with isolation are some of the factors attributable to this long existing system of farming. The adverse effects caused by practising this system are that the large scale deforestation at the hill-tops deprives them of the water source and brings a number of undesirable changes due to the disturbed ecosystems. The other obvious influences are the severe soil erosion which leads to low fertility, increases the intensity of floods by the deposition of silt in the streams. This system of farming is effectively practical with the age-old technology resulting in low productivity. There are a number of other indirect factors associated with this system which are responsible for the slow development process. The people engaged in the system, by and large, may not settle permanently at one place, this does not permit one to develop readily the basic facilities for education, health and communication. These factors contribute to their isolation and they have hardly any feeling of modernization. Schlippe (1936) indicated that the modern methods in agriculture for improving the farming systems in shifting cultivation areas could not be employed as they were incomplete. He suggested that the methods which could be employed without impairment to the existing framework of the tradition and environment. The basic factors for failure to attract the farmers engaged in shifting cultivation to permanent settlement as indicated by Borthakur et al. (1979) are as follows :

1. The new settlement disturbs abruptly their socio-cultural life.
2. Modern technology for the development of terraces, water management and water conservation are inadequately developed for this system.
3. The removal of top soil for developing the terraces causes low production in the first year.
4. The jhumias (farmers practising shift cultivation or 'Jhuming') are not accustomed to bullocks and other requisite implements for cultivation in terraces.

They suggested partial bench terracing on one-third of the slope on the foothills and growing horti-pastoral crops on the above; two-thirds area of the slope may be used as immediate alternative to shift cultivation. This statement has been supported by the quantitative data and the economics has been worked out. It is indicated that the net income realised from one hectare was highest about Rs. 2577 in the case of the three-tier unit, followed by Rs. 2373 in the complete terrace unit as against Rs. 671 in the Jhum unit.

The coastal and fresh water systems

The coastal wet lands and shallows like estuaries and mangrove swamps are sources of food for fishes, crustaceans and molluses. Although the water fowl, coral reefs are of local importance, these are potent sources of providing habitats for the fish on which many coastal communities particularly in developing countries are dependent. These play a vital role in building organs which maintain the islands. It is most unfortunate that many flood plains, wet lands and coral reefs are being destroyed thoughtlessly e.g. in Sri Lanka, the corals were extensively removed for the production of lime. They caused irrevocable damage to the fisheries, coconut groves and small lagoons. Large scale destruction of fisheries has been caused by the construction of dams and other developments like housing, roads and airports. The construction of dams blocks the passage of migrating fish and destroys the habitats for others.

Agricultural systems

"Potentially arable land" is limited to 11% of the world's land. Even this is not evenly distributed : Europe has 36%, Central America 25%, North America 22%, South Asia 18% whereas South-East Asia has 14%. A considerable area of potentially arable land is being used by buildings.

It is estimated that with the current rate of land use, nearly one-third of the arable land will be destroyed in the next 20 years. Much of the life support system is dependent on the soil. The vegetation and soil are often not balanced particularly because of poor management. It causes increased soil erosion. This results in considerable irrevocable loss of top soil. Nature takes about 100-400 years to generate 10 mm of top soil. India alone loses annually 600 million tonnes of soil depriving the country of nearly six million tonnes of plant nutrients.

In India 44 per cent of the total land area is subjected to severe erosion and obviously needs immediate attention. About four million hectares of land are under ravines. Many reservoirs constructed under the valley projects are setting fast. There are 30 major river valley projects covering the total catchment area of nearly 77.5 million hectares in the country, about 11.6 million hectares is in the threatened areas and needs immediate conservation treatment. This damage is attributed to the negligence of human beings by way of shift cultivation on fragile soils and forest clearing in flood-prone areas. The maintenance of the agricultural ecosystem depends on the maintenance of the soil together with the beneficials—habitats, insects and other animals e.g. predators, parasites of pests and crop pollinators. Excessive chemical use promotes resistance. The number of chemical resistant insects have increased almost 100 per cent within a period of 12 years from 1968-1980 (Table 9).

This suggests that it will be more effective if these are controlled by a combination of chemical control and other methods like mechanical methods, repellents and hormones and predators, and finally pest resistant varieties. In order to breed disease and pest resistant varieties, new genetic resistance sources are of paramount importance. It is thus important to conserve the genetic variability of the crop plants particularly

TABLE 9 : **Number of pesticide-resistant species of insects and mites**

Year	No. of pesticide-resistant species of insects and mites
1965	182
1968	200
1970	228
1977	364
1980	400

of economic importance from over grazing, poor management of irrigation systems, indiscriminate felling of trees for fuel, and inadequate anti-desertification measures like green-belt planting. Realising the importance of land and water conservation, the appropriate technology for its management has been devised by the scientists of the Central Soil and Water Conservation Research and Training Institute which they demonstrated through an operational research project in Fakot Watershed in the Tehri-Garhwal region. About 95 families living in the watershed in eight hamlets are engaged in farming of subsistence nature. As a result of improved management one person in every two families could afford to move to other areas for additional earnings. The waste land is being brought under horticultural trees, fuel-cum-food plantation. Bushwood and logwood check dams have been put up to prevent soil erosion. In addition soil erosion control structures have been created to break slopes. This is an example of integrated systems which provided the food, feed and fuel without hampering the natural ecological balance. The small farm water storage tanks are useful for conservation of soil and water and generate the facility, for the raising the crop year round. It has been indicated that the combination of mechanical and biological measures are useful methods to protect hill roads economically. Mulching, contour wattling and planting of trees, shrubs and grasses are some of the biological measures to be adopted to provide protective cover to soils on the slopes.

9

Genetic Wealth and its Erosion

RAVEN (1976) has reported that there are about 300,000 species of green plants, algae in the world utilizing the energy of the sun. The survival or extinction of these plant species is associated in one way or other because of the nature of food chains and the degree of specialization in the feeding habits of these heterotrophs. The stability of the ecosystems of the world is controlled by the diversity of the plants and other organisms. The conservation of the plant world is of paramount importance for the survival of mankind. It is reported that there are roughly 240,000 species of plants on this earth, these are presumably distributed nearly in the order shown in Table 10.

TABLE 10 : Species of plants distribution

	Number of Species
A : Temperate region	*85,000*
1. North temperate zone including Northern and Central Mexico, North Africa and the Himalayas	50,000
2. Temperate to arid Australia and New Zealand	15,000
3. The Cape Region of South Africa and bordering arid zones	10,000
4. Temperate South America	10,000
B : Tropical region	*150,000*
5. Tropical and semi-tropical regions of Africa	30,000
6. Tropical Asia, including New Guinea and tropical Australia	35,000
7. South America, Central America and tropical North America	90,000

It seems that about two-thirds of the plant species of the world are of tropical distribution. Swaminathan (1979) indicated that a narrow genetic base confined to highly specialised adaptations could lead to an evolutionary blind alley and thereby to extinction. The haphazard interference of man in the ecosystems has increased the pace of extinction of species and of genotypes within species. The International Union of Conservation of Nature and Natural Resources (IUCN) has reported that an average of one species or subspecies of animal is lost each year. Nearly one thousand birds and animals are presumably faced with the danger of extinction. It has been reported by IUCN that roughly 10 percent of the flowering plants of the world are under threat of extinction. It is indicated that many plants and animals of the original tropical forests that have migrated to populate sub-tropical, temperate and colder regions will by and large, disappear if the present trends of usage continue for many decades. Possibly, about 4,500 of the total nearly 85,000 species for the entire temperate regions of the world could be considered threatened. Raven (1976) reported that according to estimates of the IUCN Threatened Plants Committee (TPC), there are about 750 of these 4,500 species which should be considered endangered or vulnerable in the strict sense.

There are about 150,000 flowering plants of the tropical regions in the world. They are faced with destruction of habitats coupled with the lack of adequate knowledge of plants. There are very few botanical gardens and that too with inadequate management. According to FAO estimates about ten million hectares of tropical forest are being annually felled. It is indicated that within twenty-five years no undisturbed tropical lowland forest in the world will exist (Anonymous, 1974). It is believed that one-third of the tropical plant species will head towards extinction by the year 2,000 with the first major wave of extinction in the South East Asia and New Guinea within the next decade.

The formidable task of preservation of genetic wealth is of paramount importance for improving agricultural, forestry and fisheries production. The usefulness of these materials for the future options as raw materials for scientific and industrial innovations is of utmost importance. It is beyond our reach to predict what genetic materials may become useful. It is estimated that there are 3 to 10 million of plant and animal species. Of these only 1.5 million species have been reported scientifically. There are a limited number of species which are considered to be of economic value; these constitute wheat, rice, maize, millet, sorghum, barley, rye, potato, sweet potato, cassava, yam, common bean, soyabean, groundnut and banana. These plant species provide as nearly 90 percent of all the food requirements in the developing countries. Generally the varieties cereals like wheat do not last beyond 10 years. At times these varieties lose acceptability within a short time. The replacement of these varieties is necessary because of the change of the consumer; climate changes, pests and diseases may evolve new strains and overcome the resistance. Serious food shortages will result if such important crops are affected by either pest, epidemic or weather alterations. The dwarf habit in wheat and paddy has led to abundant increased yields in several parts of the world. There are a number of notable events in the history of plant species which have faced major drops in food output due to the outbreak of specific pests; some of these are listed below :

1. Irish famine of the 1840s caused damage to the potato crop by the late blight caused by *Phytophthora infestans*.
2. The severe losses caused by rust epidemics in wheat in North America in 1917.
3. The Bengal famine in India in 1943 due to the damage to the rice crop by the brown spot disease caused by *Cochliobolus miyabeamus*.

The production of crops from monoculture may in large contiguous areas of the same genetic strain of major crops may lead to disastrous results. Many remarkable technology advances in yield or quality in economically important plants has stemmed from one or few genes. Such commercial crops pose the danger of serious loss due to ravages of diseases or pests. This kind of genetic vulnerability was exemplified by the Victoria blight epidemic caused by *Helminthosporium victoriae* during the 1940s on U.S. improved oat varieties. The crown rust-resistant Victoria variety was used as common parent in these improved oat varieties. Subsequently, it was noticed that the gene Pc-2 for the Victoria type of rust resistance was linked with the gene Hv for susceptibility to the blight.

Another instance of far greater magnitude of genetic vulnerability to pest epidemics was marked by the Southern Corn Blight caused by *Cochliobolus haterostrophus* during 1970-71 on all the United States corn hybrids carrying T-type cytoplasmic male sterility. A similar epidemic was witnessed in the large areas in the Philippines and in Indonesia planted with a few semi-dwarf rice varieties during 1974-76 by rapid shift from the rice brown plant hopper (*Nilaparvata lugens*) biotype 1 to biotype 2. The brown rust infection to wheat variety 'Kalyan Sona' widely planted in North India during 1973-74 suggests the need for varietal and genotypic diversity.

A similar situation prevails in many other crops due to the narrowing genetic base of the commercially important crops. 75% of the wheat production is based on four varieties in the Canadian prairie, and more than half is devoted to a single variety. 72% of the potato crop grown in the United States depends on four varieties and pea production depends on two varieties. Harlan (1972) reported much of the coffee trees in Brazil descend from a single plant and the soyabean industry in the entire U.S. is derived from six plants from one place in Asia. Frankel (1973) suggested that the alarming situation caused by the narrow genetic base in crops such as wheat, rice, millets, beans, tomatoes, potatoes, banana, lime and oranges could be taken care by gathering the range of existing genetic variation in the wild and domestic types. The replacement of the traditional varieties by the new ones is an essential process to meet the requirements. It is urged that the wild relatives and the traditional varieties should be preserved to meet the unforeseeable needs of the future. This can be noticed by the epidemic caused in the grape vine in Europe by the pest *Phylloxera* in the 1860s. It affected almost every vineyard. The wine production of Europe could only be saved by grafting the European vine on American rootstock found to be tolerant to *Phylloxera*.

The importance of the wild relatives does not need to

account for the production of food, fibre, drugs and for advancing the scientific and industrial innovations. The wild life in both developed and developing countries is a subsistence and recreational source. The rural communities in developing countries depend on the wild plants and animals as renewal sources of food. These wild plants and animals are generally ignored by nutritionists and are not included in their lists as these are found in the areas which are far from the scrutiny of statisticians. Over 60% of animal fats consumed is derived from wild animals in parts of Ghana, Zaire and other Western and Central African countries. Wild animals and plants provide a significant source of income particularly for rural communities. King (1978) has reported that in 1975 the USA alone imported more than $1,000 million of wild life products. There are about 40,000 trappers engaged in the fur trade mostly from fox, beaver, muskrat, lynx and sealmink. A number of wild plants and animals are used for deriving the products for perfumes, cosmetics, pharmaceuticals, decoration, specimens for natural history, leather, skin, meat, fish etc. and for live plants and animals. In Kenya, tourism primarily based on wild life is among the top three foreign exchange earnings. There are a varied number of persons engaged in one way or the other in wild life, like hunting, fishing, licence holders, photographers and nature hikers.

10

Priority Requirements

IT HAS BEEN estimated that plant species varying from 50 to 200 become extinct annually; presumably the proportion will gradually increase. These include the plants for which the world will become poorer. Immediate attention needs to be paid to this imminent problem. It is generally considered that the species endangered throughout their range and the species that are sole representatives of their family or species deserve priority, now, else next week will be too late for at least one plant, somewhere. Looking at the large number, the priorities at national and international level have to be fixed for this formidable task of conservation. It is considered that the greater the genetic loss, the less imminent that loss need be to justify preventive action (Fig. 18).

Size of loss

The extent of potential genetic loss is related to the taxonomic hierarchy. The different positions in this hierarchy tend to reflect the genetic difference and hence difference in such variables as morphology, physiology, chemistry and ecology. The extent of difference between genera and between species within genera varies. The current taxonomic hierarchy is the only means for providing the degree of potential loss of genetic material. The monotypic families and genera should receive priority over the polytypic ones.

The smaller the size of family, the greater the magnitude of difference expected between the nearest related family; and similarly, one group of species will be more distinct from the species of the other group. It is therefore suggested that an endangered species should be given priority over a vulnerable,

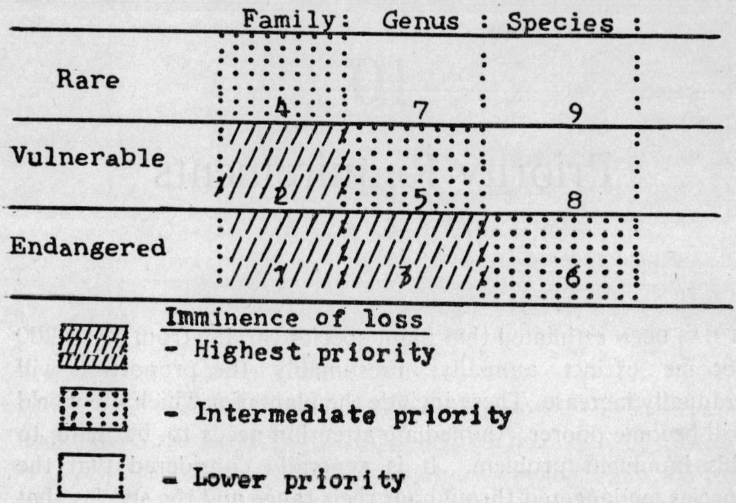

(1 to 9 indicates order of priority)

FIG. 18 : *Formulation of priority requirements of species*

vulnerable over a rare species, and rare species over one insufficiently threatened to qualify for one of the three types. The precise extent of threat is a matter of state of knowledge but with highly restricted distribution e.g. *Citrus indica* should be immediately protected. The species with highly restricted distribution need a close watch, and particular attention should be paid to families and genera. Melville (1970) reported that the Survival Service Commission of I.U.C.N. intended in 1968 to prepare Red Data Books for flowering plants and for gymnosperms for a precise and clear understanding of endangered species as a first step towards a rational programme of conservation. He indicated that the S.S.C. has already published Red Data Books for mammals and birds and the first part of the plant book in loose leaf was about to be published. A separate page has been allocated to each species thus a provision has been made to revise and replace the pages, if necessary. The provision for inserting fresh pages for additional plant species has also been made. There is a prescribed proforma as indicated in Table 11 for including a species in Red Data Book.

TABLE 11 : **Proforma indicating the IUCN Plant Red Data Book**

1. Status
2. Distribution
3. Habitat and ecology
4. Conservation measures taken
5. Conservation measures proposed
6. Biology and potential value
7. Cultivation
8. Distribution
9. References

These simple proformas need precise information pertaining to the species. These proformas have been suggested for collecting information in several ways by different countries. Jain (1981) has indicated that the following data are required for a threatened species :

1. Correct taxonomy and description.
2. Habitat and climatic requirements.
3. Known number and size of populations.
4. Whether known to be grown under cultivation in any botanic garden, methods of cultivation, etc.
5. History of the species, i.e. when first described, expansion of distribution, any known serious destruction, species competition (specially with exotics), trends of population etc.
6. Pollinators, methods of reproduction.
7. Known or potential economic uses/medicinal uses.
8. Any known national or international trade.
9. Why considered threatened ? Not collected for long, population shrinking etc.
10. Nature of threat, natural man made, biological, etc.
11. Suggestions for conservation.

A set of conservation categories of threat which are parallel to those used for animals are as follows :

i. Endangered : in danger of extinction. Survival unlikely

without protective measures.

2. Vulnerable : not yet endangered but likely to be endangered if the causal factor continues operating.
3. Rare : threat of extinction but not yet endangered or vulnerable, world population is small in specialized habitats that it would be possible to endanger them or even make them extinct e.g. by a sudden change in land use in one area.

Although the uncommon and inaccessible plants may or may not be faced with the danger of extinction, they deserve to be preserved because they may have potential, food, medicinal or some other value. This has been exemplified recently by realising the importance of periwinkle (Catheranthus rosus, syn. Vinca rosea) widely grown for ornament and found to contain 50 different alkaloids, at least two of which are valuable drugs. There are many plant species whose economic value is yet to be realized and tapped. The U.S. National Academy of Sciences has made a compilation of a few such species. Realising the importance of the underutilized, an international council for their research has come into existence. Several wild edible plants from the Himalayan region have been reported by Singh and Arora (1978). Some of these plants like *Chenopodium album* possess great potential for phytomass production. The natural rubber derived from *Hevea brasiliensis* has become economically important due to the increase in the price of petroleum products. There are annual plants like *Guayule* (Parthenium argentatum) whose latex is comparable with the product obtained from *Hevea*. Melville (1970) reports that the Cooktown Orchid, *Dendrobium bigibbum* which is the State flower of Queensland has a scattered distribution throughout the Cape York peninsula. It has disappeared from more accessible places after it was gazetted. Commercial propagation seems to be the solution to meet the requisite demand. It has been reported that there are number of stray cases where species have been rescued by the accident of cultivation. A good example of it is the Maiden-hair tree, *Ginkgo biloba* which would have been known as a fossil if it had not been discovered in Chinese temple gardens. The Dawn Redwood, *Metasequoia* is another example which would have been known as a fossil if its last stand was not discovered in China. Presently it is abundant under cultivation. The Monterey Pine, *Pinus radiata* is one of the important sources of softwood timber in the world and is being wild cultivated. It was reduced to two small populations in California where it appeared to be on the verge of extinction. There are some species like the common horse chestnut, *Aesculus hippocastanum* which finds survival difficult as it has confined to one restricted area in Northern Greece and adjacent parts of Albania. The flowering tree Flamboyant, *Delonix regia* so widely cultivated throughout the tropics is reduced to a single native colony in one limestone gorge in its native Madagascar. A number of other plant species faced with danger of extinction have been saved by cultivation during this century. The species, *Camellia granthamiana* and *C. crapnelliana* were known as from single tree in Hongkong island. The *C. granthamiana* has the largest flowers in the genus and is not now widely cultivated. However the *C. crepnelliana* with its unique characters seems to have been extinct. There are at least a few cases where there is enormous variability in a species which might prove detrimental to its maintenance of characteristics. The individual with larger flower or fascinating vigour may be extensively removed. A classical example of this is some of the Mediterranean *Ophrys* in some species of which no two individuals may precisely resemble each other.

The goats in the three Kings Islands lying off the North Coast of New Zealand reduced a number of species to the verge of extinction before they were all shot in 1946. It was possible for the Three Kings Cabbage tree, *Cordyline kaspar* to make recovery. It is a handsome small flower tree and suitable to mild climates wherein it could well replace *C. australis*. The Three Kings Trumpet-flower, *Tecomanthe speciosa* was reduced to a single plant. The beautiful vines appeared when the trusses of creamy white trumpets were displayed against glossy dark green foliage. It was initially propagated by cuttings but it can be regenerated from seeds.

A considerable proportion of the South African *Proteaceae* including *Orothamnus zeyheri* and a number of other genera including *Protea* are threatened. The Blushing Bride *Serruria florida* is used in wedding bouquets has been brought into cultivation. There are a number of South Africa Ericas which require protection. These beautiful plants can be grown with the aid of a green house. *Erica hibbertia* was named after George Hibbert of Clapham who was one of the Erica enthusiasts. It was introduced into Britain in 1800. The *E. chrysocodon* Golden Bell Health and *E. hibbertia* are from the Franschhoek Pass area favouring marshes and rocks. The *E. junonis*, Juno Health has the largest flowers of the genus and so rare that it has only been possible to collect four times since 1891. The morphology of the flower is the lack of specialized pollinators i.e. flies and birds are also important causes of its being extinct.

An intensive study is required for the clear understanding of the causes of threat. This involves the keen observations pertaining to the biology susceptibility of species to restrict their regeneration through species competition, absence of pollinators, plant pathogens, insect attacks, inadequate reproductive mechanism etc.

Conservation requires proper planning, financial resources, technical personnel, allocation and management of land and water uses etc. It is highly useful to preserve as many cultivars as possible of crop plants, forest trees, livestock, animals and domesticated organisms and their related types. Since the number of varieties is too large, priority should be given to those varieties that are most needed for our national and international programmes, the threatened types including their wild relatives deserve equal attention. Effective conservation requires both off site and on site preservation and the possible collaboration of national and international programmes for the preservation of genetic resources.

There are broadly two ways of preservation of genetic diversity.

OFF SITE

(1) *Part of the organism*

This involves the preservation of seed, semen or other element from which the organism concerned can be reproduced.

(2) *Whole organism* (*ex situ*)

The stock of the individual organism concerned is

kept out of its natural habitat in plantations, botanical gardens, zoos, aquarium, ranches or culture collections.

ON SITE (*in situ*)

In on site preservation the stock is preserved by protecting the ecosystem in which it occurs naturally.

All collections whether seed or vegetative must be duplicated under the conditions of long-term preservation. The duplication of vegetatively maintained collections needs particular attention.

OFF SITE PRESERVATION

This involves the following important steps :

1. Exploration of available diversity of target species.
2. While collecting priority should be given to the materials likely to be lost.
3. Preservation in storage in case of part of the organism.
4. Or maintenance in a plantation, culture collection etc. in case of whole organism.
5. Documentation—The systematic description giving details of place or orgin, taxonomic and morphology features should be recorded.
6. Recording of conspicuous agronomic characteristics.
7. Recording of yield potential.
8. Recording of nutritional qualities.
9. Recording of biotic qualities like pest and disease resistance.
10. Recording of eco-edaphic properties like drought and temperature tolerance, responsiveness to varied soil conditions.
11. Utilization.

Conservation is not just seed storage. The germplasm should be collected first and then be properly evaluated. The usefulness of the gene-bank lies if these being fed with the proper germplasm. It is urged to establish long-term seed storage facility to serve as a base collection. India plans to create such types of storage facilities in the near future for the region as a whole. Free access to germplasm and information held in the participating countries in principle, is accepted (Anonymous, 1978). In case any participating country desires that in some cases it would wish to exercise a degree of restriction for particular plant species, it should notify the IBPGR. Environmental controlled seed storage facility for conservation of active collections at national level is intended.

A germplasm repository in the National Bureau of Plant Genetic Resources at New Delhi with a storage capacity of about 2 lakh seed samples under long-term and about 4 lakh sample in the medium-term storage is being built.

Among the several threatened plant species, some important medicinal plants i.e. *Dioscorea* spp. *Epherdra*, *Aconitum*, *Rauwolfia*, in India and Nepal and wild relatives of sugarcane in north-east India have been included. Sugarcane, some horticultural plants like *Citrus* spp., banana, mango and medicinal plants are among the vegetatively propagated crops that deserve high priority. Some of the widely growing species are faced with the danger of extinction due to over-exploitation. The international trade in cacti from Mexico has become a matter of concern due to the depletion of their natural populations. Nearly five lakh plants of cacti are removed in Texas alone each year.

The appropriate collections representing the germplasm of the region should be established in the South Asian countries. This will be particularly useful in view of the essential quarantine barriers which may restrict the transfer of vegetative material between countries. The maintenance of genetic stocks which are prone to virus attack requiring maintenance under virus-free conditions elsewhere will only be possible in the event of active cooperation of the different countries. A good example is world sugarcane collections at Cannanore (Kerala State) where stocks are kept free from virus mosaic. There is mounting concern for establishing a virus-free site for the *Citrus* material to be collected in Sri Lanka.

Off site (*ex-situ*)

There is a long history for living collections both in plants—in botanical gardens and animals—in zoological gardens. It is felt that IUCN has in the past devoted considerable efforts to the role of zoological gardens in collecting conservation aims. The experience gained with animals including its impact on education and awareness and by generating more support for conservation is of great interest to the application of botanical gardens. However, zoological gardens have been accused of being "consumers" of wild species as they are prepared to pay large sums of money to acquire unique species. There is every likelihood of incurring similar criticism when it comes to living plant collections.

In addition to biological management, there are other important aspects like administrative, political, economic, social, humanitarian and sentimental, some of which may be a part of the very culture of the people, which deserve special attention. It can be exemplified by the religious sentiment regarding monkeys in India. The question of direct control of monkeys does not arise. The other classical example is 'nilgai', that bovine-like antelope, which because of its resemblance to the cow, is not shot. Similarly, there are a number of trees and plants like pipal, banyan, 'tulsi' which deserve mention as these are attached to religious sentiments.

There exist a series of energy transfer of food chains, from plant eating animals which obtain their energy from plants and pass on the same through the meat eaters which prey on them, up to the top-predator. The elimination of any one species causes a break in the food chain. The basic consideration should be to prevent disappearance of native species both plants and animals, preservation through wise use. The number of Great Indian Rhinoceros, was reduced to about a dozen individuals in its natural home in Assam, the Kaziranga Sanctuary. Similarly, within the last century the Indian lion has dwindled to the extent that it is confined to small area in Gir forests of Saurashtra. It has been reported that the destruction of a considerable number of pythons and lizards in the recent past in Assam had resulted in rats increasing to such an extent as to damage plantations. Similarly in South India, the reckless destruction of snakes and lizards for their skins resulted in plagues of rats (Stracey, 1966). The haphazard destruction of pig and deer, which are the food of larger carnivorous animals encourages the production of man eaters and cattle lifters, whereas the destruction of the tiger and leopard results in increase of pig and deer number which causes damage to plantations. Precisely, a similar situation exists among plants where the indiscriminate destruction of flora results in the extinction of valuable genetic diversity.

The wild relatives of plants are mostly confined to the forests, so are the wild animals which are supported and protected by the forests. The forests need wild animals, though not to the extent that wild animals need forests. The birds that eat the insects in a way assist the reproduction in maintaining the health of the trees in the forests. They also assist in reproduction and spread of certain plant species by eating the fruit and expelling the seed. In addition the soils are improved by animal burrowing, droppings and remains. Both plants and the wild life are benefited by the growing of mixed forest crops, controlled grazing and proper use of fire in a forest.

The botanical gardens make dedicated efforts for conserving endangered plant species by collecting seeds or viable parts of the plant which help them to reproduce under controlled conditions. The gardens started as a space around dwellings where nourshing and useful plants were grown. In cases where it was felt necessary, the required plants were brought from farther off places. This was admittedly the first sort of conservation. In earlier days, in mediaeval times, the varied types of plants were grown together in hedged or fenced enclosures and subsequently in walled gardens. The people growing them were called gardeners and herbalists. By and large, the persons engaged in the systematic study of plants were called botanists and became scientists. As a result, the several useful plants from ancient times still survive like orange, olive and Madonna lily. The efforts of the early conservationists has made it possible for France to maintain ancient pears and roses; England to maintain ancient cultivars which illustrate the development of today's apple in cultivation; India to maintain the ancient *Citrus*, mango and banana cultivars. Some of the criticisms to which this system is exposed are as follows:

1. The environment of the botanical gardens and natural habitat may not be alike. Consequently collections

reproduced in a botanical garden may differ from that of the natural habitat.

2. In botanical gardens, generally small isolated populations of a single species are dealt with, not the biotype, hence the simulated environment cannot restore the whole spectrum of the ecological condition.
3. The eventual reintroduction into the wild may become problematic.
4. There are some limitations that must be carefully viewed before their introduction into the botanical gardens. These introductions may bring certain pests with them or become susceptible to certain pests.
5. The explorer may possibly collect the last species or the population of the species may dwindle considerably from the last area where they are found wild. This may sometimes generate resentment from the local people, or competing institutions.
6. Collections of unique specimens from different countries (*Hevea* from Brazil) may be resented and entry permits for future expeditions may be refused.

Botanical gardens play a paramount role in the conservation of plant species as the following will show :

1. The botanical gardens can effectively save the plant species from extinction particularly, rare types. In Britain, the sedge *Schoenus ferrugineus* dwindled due to construction of a reservoir in Scotland, and it survives as a single clump in the Cambridge Botanical Gardens.
2. The wild species being preserved in their natural habitat should also be preserved off site in botanical gardens as a safeguard against any failure in on site preservation.
3. It seems easier to produce large populations of individuals of one plant species in comparison to wild animals in a relatively short time.
4. The endangered established collections in the botanical gardens can be used to re-establish them in their

natural habitat, once the danger that caused their depletion has been overcome.

5. These are helpful for building up of the structure to effectively conserve and manage useful unique resources for national and international programmes.
6. Every possible effort should be made to conserve these collections by providing them the conditions similar to those of their natural environment so as to allow the same line of evolution as in nature.
7. The genetic diversity should be preserved by keeping the populations at a stage ofev olution at which they were taken from nature. In order to represent the gene pool, sufficient large populations should be grown for the maintenance of genetic variability.
8. These should assemble their own speciality for basic and applied research.
9. They should preserve plants whose economic potential has not been properly investigated.
10. These collections should be properly recorded. The details of recording is inadequate in many botanical gardens. Even in botanical gardens like the University Botanic Gardens at Cambridge, it is only very recently that proper attention for its accession and cultivation index is being paid. These details of the stocks have made it possible to conclude that there are fifty nationally rare species of known wild British origin at present grown in the Garden. Walter (1973) indicates that a policy has been made at Kew for replacing British species in cultivation with materials of known wild British provenance. Obviously, the botanical gardens with adequate accessions and indexing system can play a paramount role in preserving the threatened species.
11. The National Botanic Garden, Howrah, preserves a number of rare plant species. The banyan tree known for distinguished characteristics is one of its specialities.
12. These should be increased for distribution also.
13. Distribution, apart from the delight of sharing is the surest way of preservation.
14. Plant species should be used for education purposes.

15. These gardens should provide facilities for research and training.
16. These should be used to generate public opinion for conservation.
17. These should become the recreation spots for the people.
18. They should attract tourists. It becomes therefore essential that there should be proper approach roads and the basic amenities for refreshment etc.

Private gardens have played a very important role in the preservation particularly of rare plants though often not deliberately. There are a variety of tastes all over the world which urge the necessity of acquiring rare plants. Every garden irrespective of its size, is a potential source of a rare type. The National Trust, the largest private garden in the world seeks to foster collections started in separate gardens and enlarges upon them. The Trust has a varied number of collections of magnolias, rhododendrons, camellias, ancient European roses, *Hypericum*, *Deutzia*, *Hydrangea*, *Acer*, *Sorbus*, Yucca and a variety of *Convallaria majalis* and *Anemone nemorosa*. In addition there are numerous autumn colouring trees and shrubs. The National Trust is preparing a list of all the major plants in the gardens. The Garden History Society is compiling a list of scarce and important plants. It is hoped that it would be desirable if each organisation such as horticultural institution, public park, botanical garden, national trust garden etc. should take on one or more genera (Thomas, 1976).

11

On Site (*in situ*) Preservation

FOR A FAIR deal, the subject on *in situ* preservation of genetic diversity deserves special attention. It is admittedly the most ideal measure for preservation. However, this method is not suitable for domesticated species. Particular attention should be paid to the preservation of the natural habitats of endangered and unique genetic material, the wild relatives of economically valuable and other useful plants and animals and unique ecosystems. The genetic base of forestry and fisheries should be preserved to make as complete a transition to domestication as crop production has been done.

The first preservation in the sal forests of Bihar (Saranda forests) was established in 1906 in India. Barring these sporadic studies, in India the selection of the representative sites for the preservation of the rare trees started only after the recommendation of the Third Silvicultural Conference held in 1929. Subsequently it was viewed that the selected sites should be preserved possibly in the present form thus the preservation of both climatic and seral types was envisaged. The recent studies of the existing preservation plots of the forests made it possible to compile the information regarding their typology, conservation of environment and to prepare an up-to-date list of existing preservation plots in the country. (Anonymous, 1975). The distribution of preservation plots in different States and Union Territories both in natural forests and plantations and the area covered with these forests is indicated in Table 12. There are 188 preservation plots, 163 natural forests and 25 plantations covering an area of 8422.4 ha. The preservation area varies from 4.0 ha in Meghalaya to 4,029.0 ha in Assam. The commercially important species i.e. *Shorea robusta*, *Tectona grandis*, and conifers like *Abies pindrow*, *A. spectabilis*, *Cedrus deodara*, *Picea smithiana*, *Pinus roxburghii*, and *P. wallichiana* cover 89 plots. Although 188 preservation plots have been established, yet many forest types like, Southern tropical, Wet forest, Semi-evergreen forest, Tropical thorn forest and tropical evergreen forest need particular attention.

TABLE 12 : **Preservation plots in different States/ Union Territories and their aıeas**

Sl. No.	States/Union Territories	No of Preservation Plots			Total area of preservation Plots (ha)
		Natural Forests	Plantations	Total	
1.	Uttar Pradesh	32	—	32	308.4
2.	West Bengal	27	2	29	294.8
3.	Madhya Pradesh	21	—	21	2844.9
4.	Gujarat	17	—	17	113.2
5.	Himachal Pradesh	6	6	12	30.3
6.	Karnataka	11	—	11	136.8
7.	Maharashtra	11	—	11	415.3
8.	Tamil Nadu	2	9	11	15.0
9.	Bihar	9	—	9	44.0
10.	Assam	7	1	8	4029 0
11.	Kerala	1	7	8	7.6
12.	Andhra Pradesh	6	—	6	118.8
13.	Andaman & Nicobar Islands	4	—	4	7.4
14.	Orissa	3	—	3	38.9
15.	Rajasthan	3	—	3	4.6
16.	Meghalaya	2	—	2	4.0
17.	Arunachal Pradesh	1	—	1	9.3
	Total	163	25	188	8422.4

Source : (Anonymous, 1975.) Preservation Plots in India, Indian Forest Bulletin No. 271 (New series) Silvics, FRI Press, Dehra Dun.

If we ever ponder the reasons for the plants being rare, it will be found that some plants are old whereas the others have evolved in the recent past and yet not had the time to spread very far. Some could not spread widely due to their adaptation to unusual environment which is found only in specific places.

Generally we have little information about the precise native habitats of the various plant species and hence the grouping task becomes extremely difficult. The primulas and rhododendrons from the Himalayas are widespread in European countries. At the other end there are some species i.e. *Orothamnus* confined to limited areas in South Africa and *Elliottia* with restricted distribution in Georgia in the valley of the Savannah River which appear to have lost their power of adaptation and regeneration dwindling down the path to extinction. Recently, the Georgia Heritage Trust, a department of the Georgia Natural Resources has acquired 750 acres for growing the surviving population of *Elliottia racemosa.* This will preserve the Georgia Plume and its habitat where an in-depth study of its reproduction *in situ* will be possible. This may also enable it to grow successfully on the off-site.

There seems to be enormous problems in selecting appropriate areas suitable for preservation as reserves in the tropics because of inadequate information. A particular example is the timber tree of the Atlantic slope of Nicaragua, which seemingly represents the undescribed genus *Flacourtiaceae* which was collected once in Belize and once in Panama and has never been collected in flower. It is feared that it will be extinct within five years in this region (Raven, 1976). It is therefore suggested that the unique ecosystem should be protected as a matter of priority. In order to preserve the range of variation particularly the different representative ecosystems rich in species should be protected. There are broadly four types of concentration of genetic resource which need protection :

1. Concentration of threatened species.
2. Concentration of wild relatives of species of economic or useful value.
3. Ecosystems of exceptional diversity.
4. Ecosystems regardless of their diversity that are poorly represented or unrepresented should be given priority.

The International Board for Plant Genetics Resources (IBPGR) has determined the priorities for crops for the global programme including South Asia and South-East Asia. These wild relatives of many crops being the major source for pest and disease resistance form the basis of the priorities. The ten high priority regions for the collection and preservation of crop genetic resources identified by the IBPGR are (Anonymous, 1980): Mediterranean; West Africa; Ethiopia; South West Asia; South Asia; South East Asia; Central Asia; Mexico and Central America; the Andean zone and Brazil. It is obvious that these are also the priority regions for selecting protected areas.

It has been envisaged at national level in India to protect the several endangered and rare crop plants of the country. These will be preserved either in the native habitat or in botanic gardens. The Government of Meghalaya has established 'Pitcher Plant Sanctuary' in Khasi Hills, at the suggestion of the Botanical Survey of India. Recently some of the Himalayan orchids and *Dioscorea* species which were facing extinction have been collected and preserved (Swaminathan, 1979). Similarly, the other threatened species growing in pure populations or in associations with other vegetation may be preserved in natural habitat e.g. sanctuary of rhododendrons in Sikkim.

12

Concept of the Establishment of the First Gene Sanctuary in India for Citrus in Garo Hills

THE NATURAL diversity for many crops prevails in forest i.e. for *Citrus*, *Musa* and *Mangifera* in the north-eastern Himalayan region. It was proposed to plan systematic studies for preserving the pockets of such habitats where considerable genetic variability occurs in economically important plants. It was proposed to preserve the important materials by way of gene sanctuaries. The gene sanctuary is a mechanism to preserve the genetic diversity of endangered species by protecting the ecosystem in which it occurs naturally. Hence it was envisaged to preserve the pockets of habitat where considerable genetic variability occurs in the endangered species of *Citrus* e.g. *Citrus indica*. The gene sanctuary for *Citrus indica* in Meghalaya has been established recently within the overall concept of biosphere reserves. The biosphere reserves in general should have the following features (Anonymous, 1977):

1. Biosphere reserves will be protected areas of land and coastal environments. They will also constitute a world-wide network linked by international understanding on purposes, standards and exchange of scientific information.
2. The network of biosphere reserve should include significant biomes and their sub-divisions throughout the world.

3. Each biosphere reserve will constitute one or more of the following categories :

   (i) Representative examples of natural biomes.
   (ii) Areas with unusual natural features or unique communities of exceptional interest. It is recognized that representative areas may constitute unique feature like,
      (a) one population of a globally rare species; and
      (b) their representativeness and uniqueness may both be characteristics of an area.
   (iii) Areas of harmonious landscape resulting from traditional patterns of land-use.
   (iv) Areas of modified or degraded ecosystem capable of being restored to more natural conditions.

   It was stressed that each biosphere reserve will normally have a non-manipulative core area, in combination with areas in which baseline measurements, experimental and manipulative research, education and training can be carried out. These areas can be associated in cluster in case they are not contiguous.

4. Each biosphere reserve should be large enough to be an effective conservation unit, and to accommodate different uses with ease.
5. These should provide opportunity for ecological research, education and training. They will have considerable value as benchmarks or standards for measurement of long-term changes on the biosphere as a whole. Their existence may be useful to other projects in the Man and Biosphere (MAB) programmes.
6. A biosphere reserve must have long-term legal protection.
7. In some cases biosphere reserves may coincide with, or incorporate, existing or proposed protected areas, such as national parks, sanctuaries or nature reserves.

It is felt that this concept of biosphere reserves is sufficiently flexible and will continue to evolve with the time. The very concept of gene sanctuary made it possible to conceive the establishment of Citrus gene sanctuary. It is recognised that the north-eastern Himalayan region is considered as the natural home of many citrus species. It has been stated unequivocally that *Citrus indica*, *C. assamensis*, *C. latipes*, *C. ichangensis* and *C. macroptera* are native of north-eastern Himalayan region. Recently the renowned national herbaria of Botanical Survey of India at Howrah, Shillong and Dehra Dun and of Forest Research Institute, Dehra Dun were studied by me to review the past work pertained to *Citrus*. These studies revealed the presence of more than 23 Citrus species in India (Tables 13 and 14). It is notable that many of these species, including the *Citrus indica* are confined to the north-eastern region. The wild types are relatively more common in the north-east and in the foothills of the Himalayan region. All except *Citrus indica* are prevalent and hence do not need any particular attention for their protection. The *Citrus indica* being the most primitive and perhaps the progenitor deserves special consideration. The distribution of various prevalent wild types in this region has been indicated in Table 15.

TABLE 13 : **Distribution pattern of 23* *Citrus* species in India based on the observations made on the available herbaria at Shillong and Howrah (Botanical Survey of India), Dehra Dun (Botanical Survey of India and Forest Research Institute) and personal survey**

State/Union Territories	Number of collections studied (Legible sheets)
1	2
Assam	38
Meghalaya	26
Arunachal Pradesh	10
Nagaland	5
Tripura	3
Mizoram	1
Manipur	1
Sikkim	1
West Bengal	14

(*contd.*)

TABLE 13—*contd.*

1	2
Uttar Pradesh	31
Orissa	5
Punjab	5
Tamil Nadu	4
Bihar	2
Karnataka	2
Andhra Pradesh	1
Maharashtra	1

*List of 23 *Citrus* species.

C. indica, *C. latipes*, *C. hystrix*, *C. decumana*, *C. maxima*, *C. gumbhiri*, *C. sinensis*, *C. aurantifolia*, *C. dimorphocarpa*, *C. acida*, *C. aurantium*, *C. medica*, *C. nobilis*, *C. limonia*, *C. grandis*, *C. limetta*, *C. assamensis*, *C. megaloxycarpa*, *C. ichangensis*, *C. macroptera*, *C. vulgaris*, *C. reticulata*, *C. limon*

TABLE 14 : **Distribution pattern of *Citrus indica* in India based on the observations made on the available herbaria at Shillong and Howrah (Botanical Survey of India), Dehra Dun (Botanical Survey of India and Forest Research Institute)**

State	Number of collections studied (Legible sheets)
Meghalaya	14
Nagaland	5
Assam	4
Manipur	1
Orissa	2
Uttar Pradesh	2

TABLE 15 : **Distribution of Wild Citrus species in north-east region**

Species	Country of Origin	
1	2	
1. *C. indica*	India	Tura range in Garo Hills in Meghalaya Kaziranga forest in Assam foothills near Dimapur in Nagaland

(*contd.*)

TABLE 15—*contd.*.

1	2	
2. *C. latipes*	India	Central plateau of Khasi Hills (Shillong) Meghalaya
3. *C. ichangensis*	India	Nagaland
4. *C. assamensis*	India	Shella area– Southern slopes of Khasi Hills in Meghalaya
5. *C. macroptera*	India	Shella area—Southern slopes of Khasi Hills in Meghalaya

The first gene sanctuary in India for *Citrus* in Garo Hills within the concept of biosphere reserves will be declared as a National Park. The various steps followed for the establishment of this National Park of Citrus Gene Sanctuary are enumerated as under :

1. In-depth study of the national herbaria like Botanical Survey of India at Howrah, Shillong and Dehra Dun and Forest Research Institute, Dehra Dun. This study included the information regarding the number of *Citrus* species prevalent particularly in India, their time and places of prevalence, including elevation, climatic features, topography and other characteristics of its natural habitat. The information like local name(s) of the variety, uses and the collector's name and address.
2. The critical study of the literature of citrus and its wild relatives was made. The study pertained to the history, origin, classification, commercial uses, centres of diversity and its association in the country.
3. A systematic detailed survey was made particularly of the centres of diversity to study the extent of each citrus species. Major emphasis was placed on the wild and rare species. The study of common orange the mandarin orange particularly was unavoidable.
4. Major stress was given to the study of the natural habitats of native wild and semi-wild *Citrus* species.
5. Considering the north-eastern Himalayan region is readily endowed particularly with the indigenous *Citrus*

species therefore the dedicated efforts were made to study this region with vast and varied diverse problems. It was observed that *C. macroptera*, *C. latipes*, *C. ichangensis*, *C. assamensis* and *C. indica* were found in the semi-wild and wild state.

6. These semi-wild and wild types are found in different pockets at varied elevations. The areas of some of these were overlapping. It was observed that semi-wild species, *C. latipes*, *C. macroptera*, *C. ichangensis* and *C. assamensis* were prevalent in the several pockets.
7. Essentially the priority of preservation is based on the degree of its being faced with the danger of extinction. The *Citrus indica* is the threatened species. It possesses varied characteristics e.g. disease and pest resistance, adaptability to wider range and several others of economic importance.
8. It seems to be the most primitive and perhaps the progenitor of the *Citrus* species.
9. It is a rare species and is confined to certain pockets of the north-eastern Himalayan region which is its native place.
10. This species has been encountered in the unique area with unusual natural features of exceptional interest. A fragment of Tura range in Garo Hills in Meghalaya is the most ideally suitable site for the establishment of *in situ* Citrus Gene Sanctuary—National Park. This population of economic plants threatened with the danger of extinction is certainly a globally important species. It is confined to the forests in a wild state. The representativeness and uniqueness are both characteristics of the area.
11. An enormous natural variation of *C. indica* is observed there. The bushes with varied prolificacy with regard to the plant and fruit characteristics were observed under different conditions. The plant characteristics varying from small bush trees to the extent of climbers were encountered at this site. In order to allow the populations to follow the same line of natural evolution, on site preservation in the natural environment is essential. It is not possible to restore the whole

spectrum of ecological conditions elsewhere if these are removed. The incompleteness of environmental conditions coupled with genotypes that are submitted to continuous alterations by recombination of genetic material, selection will follow different lines from those in nature under self-created conditions. The frequency distribution of this species was recorded in its natural habitat.

12. The core area of the site is about 10265.96 hectares. It is large enough to be an effective conservation unit for *Citrus indica* and for accommodating different uses without conflict. Possibilities are explored to increase this considerably and to protect fauna also of this ecosystem.
13. Details of flora and fauna of the ecosystem were determined.
14. A sub-committee was constituted by the National Organising Committee for the ICAR Golden Jubilee to consider critically various aspects for setting the first gene sanctuary in India for *Citrus* in Garo Hills on a similar pattern of biosphere reserves. The Committee approved the proposal with the recommendations from Dr. B.P. Pal, F.R.S., Chairman of the sub-committee for getting necessary approval from the Government of Meghalaya. Dr. B.P. Pal was the Chairman of the National Committee on Environmental Planning and Coordination.
15. The forests of this site belong to 30 Akhings each with a village head called "Nokma". These are under the administrative control of the Garo District Council. The agreement between the owners and government enables these forests to be reserved as protected areas.
16. It is an undisturbed area with varied elevations and having distinguishable flora and fauna forming a notable ecosystem. A number of streams originating in this region find their way down to the plains.
17. Subsequently, this proposal was forwarded to the Government of Meghalaya to establish the first gene sanctuary in the country and declare it as a national park under section 35 of the Wild Life (Protection) Act, 1972. The connotation of the National Park is different in various countries. The sites of historical or

archaeological importance or great natural beauty are called National Parks in England; areas exclusively faunal in character are termed as national parks in Africa; whereas in America these are sites of outstanding scenic beauty and remarkable natural phenomena.

18. A Task Force was constituted by the Indian Council of Agricultural Research in 1980 under the chairmanship of the Chief Secretary, Government of Meghalaya for the establishment and development of the *Citrus* Gene Sanctuary.

    The Task Force after holding a series of meetings with the various officials, village heads or Nokmas, the owners of land arrived at definite conclusions and prepared the blue print of the proposed site. Subsequently the Government of India decided to set up *Citrus* Gene Sanctuary—National Park in Garo Hills where citrus fruits were found originally.

19. This will be provided adequate long-term legal protection. Effective legislation is required for proper conservation and development. Laws are generally not implemented properly. It may be due to the lack of understanding information in the common man. The weak penalties coupled with jurisdiction conflicts between the agencies or between central government and local do not permit the law to work effectively. The proper coordination between the various agencies involved like Indian Council of Agricultural Research; Departments of Environment, Science and Technology, Forest, Government of Meghalaya and North Eastern Hill University etc. Since the flora and fauna both need to be preserved at the present level for the maintenance of the ecosystem, it is obvious that several agencies like wild life, forestry, agriculture, horticulture etc. will co-ordinate. Legislation concerning natural resources is being strengthened in India. The present laws are inadequately defined which envisage compensation for economic loss and no provision exists for long-term damage to its individuals due to the destruction or depletion of plant species or degradation of ecosystems. In India such ecosystems are mostly in the economically

backward areas hence the enforcement of such laws will be feasible which are ecologically, economically and socially viable.

The significance and urgency which the government attaches to conservation is evident by the recent ordinance. On 28 October 1980, an ordinance for Forest (Conservation) has been promulgated by President Shri Neelam Sanjiva Reddy for forest conservation and controlling diversion of forest land for nonforestry purposes. This ordinance comes into force immediately and provides that no State Government or authority shall make, except with prior approval of the Centre an order directing that any reserve forest area or portion of it shall cease to be reserved. Nor can they issue an order to let any forest land be used for non-forest purposes.

20. This National Park will provide opportunities for research, education and training. The ecological researches will be useful for providing benchmarks or standards for long-term changes in the ecosystem as a whole. The National Park in addition to the core zone, should have a demarcated buffer zone.

Core Zone : The uses of this area should be strictly controlled without any development. The National Park has a large enough core area to be self-sustaining which is surrounded by buffer area. There should be minimum interference and it should serve as a base-line for research, education and training activities which are carefully controlled and should be non-manipulative.

Buffer Zone : The moment are limited to authorized agencies. This area should be used for research and educational purposes.

This will provide ample opportunities for ecological research. The long-term observations may be carried out to evaluate natural and man-made changes. In situations where the objectives of experimental research and conservation conflict, research may not be allowed. The provisions should be created for education and training. The evaluation process needs the integration of contributions from ecological and related sciences. It will be appropriate if such agencies like National Bureau of Plant Genetic Resources, Department of Environment, Botanical Survey of India, Zoological Survey of India, North Eastern Hill University, Forest departments, various state departments like Forest, Agriculture and Horticulture, etc. will co-ordinate actively. In addition, educational department like National Council for Education Research and Training will be highly useful for educating the people at various levels.

The co-ordinated research for the comparison of protected areas examining of the changes occurring in natural systems over long period without human intervention with those of the adjacent areas where man-made modification to the environment may be undertaken. These studies should include the comparison between ecosystems in time for proper understanding of the dynamics of this National Park. These studies will enable to understand the environmental and other biological problems and will provide practical aspects on sound basis for the conservation of natural resources management.

The initial survey should be conducted by the group of different scientists as pointed out earlier for the proper assessment of communities including in-depth flora and fauna of the entire site of the National Park. These surveys should include photographic records, including ground photographs and remote sensing which should possibly be repeated at intervals. The research activities can conveniently be grouped as continuous or frequent, periodical or occasional.

1. Continuous research activities include the study of *Citrus indica*, the climatic and hydrological variables.
2. Periodical research should include soil features, the in-depth study of the community composition of the plant and animals, including age distribution.
3. Occasional research should be stressed on ecological and genetic processes designed to provide the scientific basis for observed changes like age structure and succession, population size and gene frequencies.

A programme related to the specific nature should be developed for effective training at the post-graduate levels, including research fellows and training courses and exchange of personnel. The involvement of the local communities is a valuable means for the successful achievement of these targets. Their participation in planning, decision-making and management is indispensable. The local communities should be educated in the importance of conservation. The direct benefits of the protected area should be well thought out and brought into the limelight particularly in the adjoining rural communities i.e. the regular supply of water. The Royal Chitwan National Park in Nepal and Amboseli National Park in Kenya benefit the local people by providing them with a regular supply of water and by providing habitat for wild life which may be hunted outside the protected area.

The present site of the National Park is an unique ecosystem which may later be designated as a biosphere reserve. Most of the problems can effectively be handled by the national programmes, whereas some of the problems can be handled only at the international level. International involvement is not only needed for the study of large scale phenomena such as climate but also as a means of avoiding duplication of research. The various international co-ordinating agencies are: Unesco's Man and Biosphere (MAB) Programme and the work of the Scientific Committee of Problems of the Environment (SCOPE), and of the International Council of Scientific Unions (ICSU).

FEATURES OF THE REGION OF SITE

One of the courses in the preservation of such useful materials is through the establishment of gene sanctuaries, wherein these native types and their related species could get protected, and species differentiation may continue in them in their native environment. A case in hand is the preservation of the *Citrus* species particularly of rare types. It was but recently, in 1978-80 that systematic surveys were conducted by me in the north-eastern region. The recent investigations have revealed that many rare types of citrus exist in this region. Since total diversity cannot be achieved within a single area, it is essential to define the levels of diversity which have priority. The priority of conservation among different species differs in view of the degree to which they are threatened with extinction. In this context *Citrus indica* which is the most primitive and perhaps the progenitor type, is highly endan-gered. *C. indica* is confined to a restricted area which has a unique ecosystem. Certain pockets where this species is still available have been recently identified. It is thus essential to preserve the specific 'unique areas', which are considered to have special importance for conservation, on the model of Bio-sphere Reserve. All the wild species of *Citrus* except *Citrus indica* are prevalent in this region and hence do not need any particular attention for their protection. In this context a site in Garo Hills is essentially the most ideally suited unit for preservation of *Citrus indica* (Fig. 19). The site of the *Citrus* gene sanctuary in Garo Hills has been marked. There exists immense natural variability of *C. indica* in this part of the Tura Range.

Tura Range including its foothills, in this context is the most suitable site. It is one of the largest ranges in the Garo Hills. The Meghalaya constitutes five hill districts, East Garo, West Garo, East Khasi, West Khasi and Jaintia. Garo Hills is situated at the western end of Patkoi Range of Assam between latitudes 25°9′ and 26°1′ north and longitudes 89°49′ and 91°2′ east and forms the western region of Meghalaya. It is bounded on the north and the west by Goalpara District of Assam and on the east by the Khasi Hills District of Meghalaya. The district borders with Bangladesh on the south and partially on the west. The district consist mostly of rugged terrain fringed by small strips of plains on the northern, western and southern peripheries. Small valleys are found in between adjacent hills at many places in the hilly tracts.

The Garo Hills has an area of 8,084 square kilometres. The Garo Hills is characterised by a dense cluster of hills of low elevation with a central ridge traversing from the north-west to south-east known as the Tura Range. Many of these hills are below 600 metres in altitude except in Arbella Range where the elevation reaches 960 metres and in the Central ridge which maintains a fairly even altitude of 1200 metres. The high-est peak is Nokrek which is 1412 metres high. The Balpakram

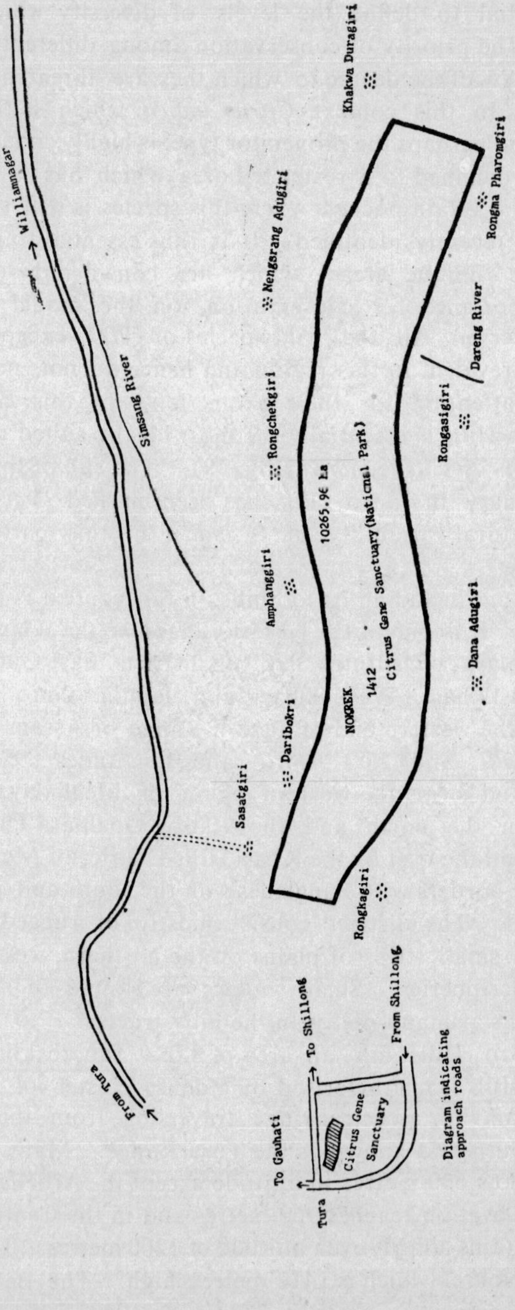

FIG. 19 : *Schematic diagram of Citrus Gene Sanctuary (National Park) in Garo Hills*

940 metres on the border of Khasi Hills and Kailas peak at 1125 metres height east of the Someswari river are also the main peaks. The Tura town headquarters of the district is situated at an elevation of 390 metres. Several rivers and streams originate in the higher hills and find their way down to the plains over narrow and rocky beds and join the Brahmaputra and Meghna basins.

Garo Hills has a tropical climate characterised by high rainfall and high humidity, generally moderate cold winter and mild summer seasons. Monsoon rains of high intensity are received during the months from April to October. The average maximum temperature in the winter months is 25°C and in summer 30.4°C. The average annual rainfall is about 3112 mm spread over 113 days. June and July are the wettest months. Generally there is a decrease in rainfall with decreasing elevations and the distance from the central ridge.

This district has a population of about 5.05 lakhs which forms about 40.2% of the total population of Meghalaya State. Over 80% of the population constitutes Scheduled Tribes where Garo is the major tribe. Other tribes are Hajongs, Koches, Ranhas, Salus and Nanais (Fig. 20). The Garos are the most important tribal community here who call themselves 'Achik' (hillman), 'Mande' (the man) or 'Achik Mande' while others call them Garo.

The majority of Garos are Christians. Their traditional religion was animism, formerly. They believed in one Supreme Being called Salgong. A number of smaller spirits and deities also existed. The priest called Kamal used to offer sacrifices for the appeasement of these deities and spirits.

Garo society consists of five exogamous groups or clans; each again divided into many sub-groups called Maharies related to the mother's side. Each Mahari lives in a specific area called *Akhing* under the authority of a Nokma. The Nokma acts as the head of the village and has the authority to settle disputes. The Garos are matrilineal in descent, succession and inheritance. They follow exogamy and hence intermarriage within the same clan is taboo. The prevailing practice of marriage is by negotiation. The marriage between a son and daughter of the same father is permitted provided their mothers belong to separate clans.

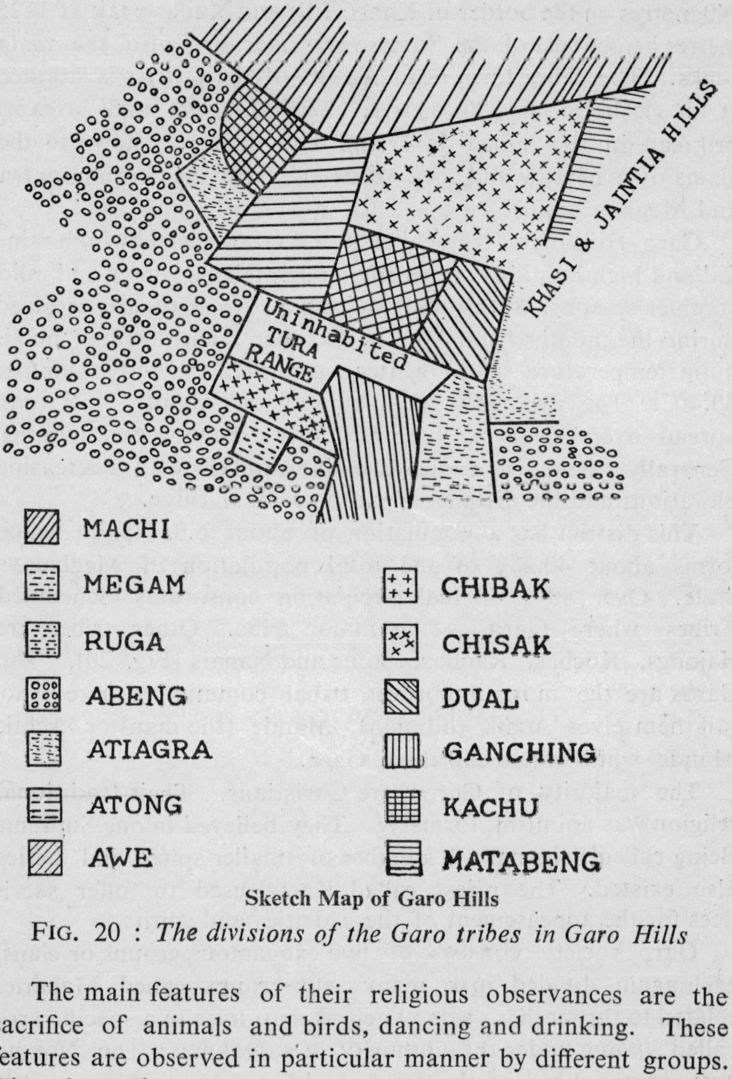

Sketch Map of Garo Hills

FIG. 20 : *The divisions of the Garo tribes in Garo Hills*

The main features of their religious observances are the sacrifice of animals and birds, dancing and drinking. These features are observed in particular manner by different groups. The clans like Akawe and Abeng have a common tradition of fixing a number of long bamboos with leaves placed upright in the ground in the vicinity of their houses. These are hung with cotton bolls or bunches of paddy in ear. These offerings when served to the dead are called memong midong and when served to spirits are called mite-midong. When an offering is made at the sacrificial altars, the blood of the bird or animal is smeared over it and some feathers are stuck to the bamboo. Pigs or even cattle are offered in sacrifice for the recovery of patients in some instances. They also observe certain ceremonies collectively once a year for the safeguard of their members from danger of sickness and all other unforeseeable mishaps.

The livelihood of the Garo is dependent on agriculture, obviously then prolonged drought or unseasonable and excessive rains causes them anxiety. According to the Garo mythology, the spirits governing the seasons are of paramount importance and hence they offer some kind of sacrifice at all the stages of cultivation and harvesting. The Garo observe religious ceremonies strictly. At the first instance of acquisition of land he initially clears a little portion at the corner of land, this is called opata. The further clearance for cultivation depends upon the nature of his dream during the following night. In case it is an unlucky dream, he abandons the land which he proposed to cultivate and seeks another piece of land. The selection of the land is followed by clearing the jungle and getting it ready for cultivation. A ceremony called Agalmaka consisting of sacrifice of fowl is performed at this stage. Similarly, sowing of seed is also preceded by individual and collective ceremonies. The priest strikes the earth with the handle of a *dao* or *choper* and reminds the spirit that certain flowers in the forest have blossomed and thus it is a sign that it is time to sow the paddy. Similarly, as soon as the crop is ready for harvest, the first fruit is offered to the gods, by way of plucking some ears of paddy or millets which are pounded between two stones and offered on a piece of plantain stem.

Earlier they used to cremate their deads; this was done by relatives of the deceased. The expenses of this ceremony were borne by the relatives of the female clan. In general today the practice is to bury the dead in accordance with Christian rites.

The villages are sited near the water sources and forests free from elephants and where good soil for Jhuming found. Among the several common facilities in a village, there is one 'Nokpante' meaning bachelor's house. Nokpante is meant only for the male youths and is strictly out of bounds to women. They have field-huts called Jamadal at the site on the Jhum field for watching over the crops, in addition to the permanent house in the village.

The principal crops in the Garo Hills are rice, maize, pulses, rapeseed and mustard, mesta, jute and cotton. The brewed liquor prepared from rice, millet and maize plays an important part in the daily life of the Garo. The liquor is always brewed and never distilled. Fish forms an essential part of their diet. Many kinds of edible wild roots and fruit are found in the forests, which the Garo make use of particularly in times of food scarcity. Agriculture is the mainstay of the Garo. A large number of cereals, vegetables and fruits are grown due to the prevalence of varied altitude and climatic conditions. Bananas, oranges and other fruits are grown over an area of about 3000 hectares. The soil and climatic conditions are extremely suitable for oranges and bananas. There are two cropping seasons in the region, the sowing of the one season starts from March and continues up to August, while the other is the Rabi season, which begins in September and ends in December.

The principal livestock of the Garo Hills are cattle, buffaloes, pigs and poultry. There is an almost negligible number of cattle being utilized for draft power. These are commonly reared by the non-Garos in the plains. The fauna of Garo Hills is listed in Table 16, whereas the important trees in Garo Hills are listed in Table 17. The Garo Hills is rich in forest resources, 2,35,573 hectares forming 32 percent of the geographic area of the district. In this district, the forests are divided for administrative purposes as state forest under the administrative control of the forest department of Meghalaya State; and forests under the administrative control of the Garo Hills District Council.

SELECTION OF THE SITE OF FIRST NATIONAL PARK FOR CITRUS

Citrus indica is prevalent in the hilly tracts of the east of Tura town. There is a metalled road of about 30 km from Tura to Orangitok and about 4 km unconstructed passage leading to Chandigiri and to Sasatgiri village. Sasatgiri is situated at an altitude of 960 metres. This is a small but compact village with scanty inhabitants about 180. This village welcomes one with

TABLE 16 : **Fauna of Garo Hills**

Sl. No.	Group	Common name
1.	Mammals	White Browed Gibbon or Hoolock, Rhesus Monkey, Assamese Macaque, Capped Langur, Sloth Bear, Rousette Fruit Bat, Short-nosed Fruit Bat, Jungle Cat, Tiger, Leopard Cat, Barking Deer, Sambar, Malayan Giant Squirrel, Palla's Squirrel, Large Indian Civet, Indian Flying Fox, Rats, Mice
2.	Birds	Pelicans. Cormorants, Herons, Egrets, Storks, Ibis, Barbed Goose, Ducks, Teals, Kites, Eagles, Hawks, Vultures, Pheasants, Partridges, Quails, Babblers, Thrushes, Warblers
3.	Reptiles	Tortoise, Turtle, Monitor Lizard, *Skinks*, Pit-Viper, Russell's Viper
4.	Amphibia	Frogs (not edible species)
5.	Insects	Butterflies, Moths

Source : Department of Science and Technology, Govt. of India.

TABLE 17 : **Important trees in Garo Hills**

Sl. No.	Botanical name	Local name
1.	Muros lacbigata	Bola
2.	Michalia champaca	Titasopa
3.	Gmelina arborea	Gamari
4.	Amoora wallichii	Amari
5.	Shorea robusta	Sal
6.	Artocarpus chaplasha	Sam
7.	Chikrasia tabularies	Bogipana
8.	Cinnamomum	Gonsoroi
9.	Adina cordifolia	Holu
10.	Alibizzia procera	Koroi
11.	Lagerstronia flosregina	Ajhar
12.	Eugenia gambolena	Jamuk
13.	Lager stroemia parviflora	Sida
14.	Schima wallichii	Jharul Gogro
15.	Duabanga sonneratioidis	Khokan
16.	Cadrela toona	Makiasal Jatipoma
17.	Eugenia grandis	Jia
18.	Terameles mediglora	Bhlen
19.	Stereospermum chelonoides	Paroli.

Source : Techno-Economic Survey on Garo-Hills District 1975. Govt. of Meghalaya.

the *chiring* (stream in Garo) and with the fragrance of the forest having three *Citrus indica* plants scattered in this isolated area. The rear part of the village begins to take the shape of a feeble forest. There are several trees of *C. indica* which are found near the chiring and are in a scattered manner.

It was noticed that at the periphery of this area, there is a narrow passage leading to a place known as Daribokri which is at a walking distance of about 5 km. It was observed that jhuming was under practice in this area and hence this area was omitted from the site of preservation. Daribokri is bestowed with ample number of *C. indica* plants. This place is situated at an altitude of about 1194 metres. Only a few huts could be seen there. At a distance of about half a kilometre from Daribokri the forest starts. The Nokrek peak is situated at a distance of approximately 8 km from this place. The elephant beds with varied sizes were prevalent up to an altitude of about 1250 metres. The *Saccharum* species proved to be forming comfortable spongy beds for them. It is a dense forest predominated by numerous perennial vegetation. Tiger, deer, junglecat are amongst the few which are also found commonly in this forest. As we proceed towards Nokrek peak the density of the forest increases. This entire region does not have any inhabitants. Even at the peak hours of the sun one could hardly see its rays. This made the entire area dark even during the day time. Several streams run throughout this region. The entire region is highly humid and dense. Several plants of *C. indica* are encountered which were isolated in scattered form in the entire site. A few of them were found in the fruiting stage in the month of February. The height of trees varied from one metre to a little over two metres. The thorns were relatively large and prominent and plants had few branches with little spread. The fruit was very small and many seeded. It has a relatively small deep orange thin rind. Considering the area, the density of the forest and the danger of the animals, it was decided to take observations pertaining to the frequency of the *Citrus indica* at random but at possible points. It was observed that this species has a frequency from 0 to 10 per 100 sq. metres. It was hard to assign the particular reasons for its distribution pattern. It was certain that highly humid and dense shade favoured the flourishing of the species.

It was notable that the trees in the dense forest bear flowers and fruits and were more prolific than those found at the periphery. This is a convincing evidence in favour of *in situ* preservation of variability. Closely resembling climbers of this species appeared casually in the dense forest. The Nokrek is the highest peak of the entire Garo Hills. This peak is amongst the few where the Britishers used to celebrate their victory after the First World War once in a decade. This victory was celebrated by lighting bonfires on these peaks.

The area of the site and its administrative control

The proposed site of *in situ Citrus* Gene Sanctuary has been shown in Fig. 19. It depicts the entire site where the plants of the *Citrus indica* have been encountered. The altitudes of Nokrek peak and also the sites of few important rivers have been indicated. The approach lines to this site have also been shown. Due consideration has been given while demarcating its boundaries. By and large, even the small villages near the boundaries have been avoided from the site. This entire gross area consists of about 10265.96 ha. This area is under the administrative control of the District Council whereas the actual land belongs to 30 Akhings. The places Sasatgiri, Mandalgiri, Rongkhinggiri, Bandigiri, Amphanggiri, Rongchekgiri, and Nengsrang Adugiri surround the northern face of the site. At the foothills of the other face, this species is prevalent near Durabandagiri, Dana Adugiri, Rongasigiri, and extending up to Rongma Pharomgiri. These plants are frequently found in these villages near the chirings, which are the protected areas commonly known as the village reserves. This area extends from ¼ to 1 kilometre in each of these villages. The plants of *C. indica* were also observed in the graveyard of Rongkhinggiri which also forms a part of the village reserve. I therefore, strongly feel that the plants of the *C. indica* found in the village reserves should also be protected immediately. The places so called villages with one or few huts have also been omitted from the site of preservation.

Fig. 19 indicates the proposed site including the adjoining villages. Although for locating the borders of the site conveniently, the villages have been shown in Fig. 19, whereas these are not included in the net area of the site. This site forms an important fragment of the Tura Range. This includes Nokrek, the highest peak in this range. The altitudes of this site vary somewhere from 344 metres at Dana Adugiri to 1412 metres at Nokrek in this range. The net area of the Citrus gene sanctuary has been marked in Fig. 19. It is about 10265.96 hectares. The estimate of the area is based on the vertical projections whereas for our purpose, the profile area is required which is far more than the indicated one.

Tura range is the source of several rivers particularly of the two main rivers viz. Simsang in the north and Dareng in the south. The tributaries of these two main rivers form a net in the entire range. The important chirings of this range are the Narong chiring, Rarabansi chiring, Chirija chiring, Rongma chiring, Rongbang chiring and Chigrang chiring.

The boundary of the site in the north starts from nearby Daribokri and extends to Nengsrang Adugiri and to Khakwa Duragiri. The southern boundary at the foothills of this range starts near the village, Rongkagiri and extends to Dana Adugiri, Rongasigiri and finally to Rongma Pharomgiri. In the east, the boundary starts from the north of Rongma Pharomgiri and extends to nearby Khakwa Duragiri. Similarly, the boundary in the west extends from Rongkagiri to nearby Daribokri. By and large, the villages are situated at a distance of ½ to 3 km away from the boundaries. The entire site is uninhabited as shown in Fig. 20.

13

Discussions

It is but recently that a number of scientists have taken painstaking efforts to make the world aware that the large number of plant and animal species are in danger of extinction and therefore need some concrete forms of protection to ensure their survival. Modern civilization has forced man to derange the natural combination of organic life and the inorganic world in order to meet his requirements. The needs of the poor man for livelihood, the greed of the middleman for quick profits, and the demand of industry have disturbed the harmonies of nature permanently due to an ignorant disregard of the laws of nature. Even the scientists have contributed to the disappearance of several species of plants during collection activities.

Uncontrolled collections and ignorant grubbing up of the plants for commercial purposes have caused the loss of several species of orchids from some parts of India, where even the last host tree of epiphytic species were chopped down. Orchids are known for their wide range of variation. There are 17,000 species of orchids and about 1,250 of these occur in India, of which about 300 are found in Meghalaya alone. These have long lasting flowers which are mainly used for ornamental purposes. It is reported that two Lady's Slipper species, *Paphiopedilum insigne* and *Paphiopedilum venustum*, the golden-flowered *Dendrobium densiflorum*, the violet *Dendrobium nobile* and many species of *Cymbidium* are rapidly becoming scarce. About 20 species of orchids have been reported that are becoming scarce in Meghalaya. The orchid sanctuaries can be useful for the conservation of this valued material of universal concern.

It has been recently felt that the undisturbed tracts in the Dangs forests in Gujarat are bestowed with the most productive wood and deserve to be protected as biosphere reserve in the larger interests of conservation. It is a rich reservoir of a number of plant species which were not considered of economic use about 20 years ago but are being considered as the most valued today. The good examples of these species are: *Garuga pinnata* "kakad" is used for pulp and packing cases, *Ougeinia oojeinensis*, 'tanch' is used for agricultural implements, *Adina cordifolia* "haldu" and *Mitragyna parvifolia* 'kalam' are utilised in bobbins and furniture and *Lannea grandis*, "modad" and *Wrightia tinctoria*, "kudi" are used for toy making.

Similar examples exist in several parts of the world, which have contributed to the loss of many plant species e.g. in South Africa the attractive plants such as Proteas and their relations; in Africa and America, isolated species of cactus and succulents; in northern Portugal and Spain, the eradication of narcissi in vast areas, and recently in Turkey, the onslaughts on the tulip and iris species. In Afghanistan people relish bulbs and tubers as food in desert areas; in Lebanon human beings eat roasted crocus corms as we do chestnuts. These desert dwellers remove indiscriminately trees and shrubs for their use. In developed countries like the U.S.A., large quantities of fritillaries and erythroniums are used to the extent that in certain areas these flowers can now only be seen in National Parks. In several places the dwellers are interested to supplement their meagre living and are not even aware of problems arising from the indiscriminate removal of plants species.

Recent examples of ignorant collecting were found in Turkey where collections of a single species to the extent of fifty thousand elements of cyclamen tubers (Cyclamen mirabile) were made in 1973 for a Covent Garden firm. This species is found only in two or three Turkish localities. The existing laws forbidding the collection of wild plants are difficult to enforce strictly.

There is another good example where the haphazard removal of the Lady's Slipper orchid in Britain has reduced it to single wild plant. The destruction of plant species is at times accompanied by preservation e.g. in Brazilian forests, the widespread tree felling to convert the forest into grazing land was accompanied by collectors who gathered orchids, bromeliads and other ephiphytes before their host trees were destroyed.

This suggests that only purposeful organized and well planned expeditions aiming to collect crop plants should be encouraged. The collections from tourist areas, nature reserves or national parks should basically not be permitted. There are some plants in nature which having failed in the evolutionary race have become rare. A good example of this aspect is the last survivor of the prehistoric *Metasequoia glyptostroboides*; now so wildly grown in Britain which were being cut for timber by local villagers when a botanist happened on the scene. There are cases where a rare wild species might have had a special interest or value. *Rheum rhaponticum* (rhubarb) introduced to cultivation in the seventeenth century was so much hybridised and developed by selection that garden plants no longer resemble the species. The species has become rare in nature and obviously deserves reintroduction by seeds which are being multiplied and preserved alive in botanic gardens for their scientific interest and breeding potential. The uncommon and inaccessible plants, whether endangered or not that are unknown are worth preserving because they may have food potential, medicinal or other value. It has been reported that the rosy periwinkle (Catharanthus roses syn. Vinca rosea) is widely grown for ornament, contains 50 different alkaloids, at least two of which are valuable drugs. It is estimated that there are about 2.50-30 lakh species of flowering plants in the world of which 50-200 plant species become extinct every year and it is expected that this proportion will increase. This will unquestionably cause irrevocable damage to the world.

There are a number of factors that have played a role in causing plants to become rare. Undoubtedly some have adapted to the specific type of habitat which is only found in a few places. Others might have evolved recently and have not yet had time to spread very far. The primulas and rhododendrons from the Himalayas which now are commonly found in Britain are probably all of Tertiary origin. While at the same time, there are a number of ancient relict species from earlier geological epochs fighting for survival among more recent and efficient competitors. The two classic examples are the Georgia Plume, *Elliottia racemosa* and Marsh Rose of South Africa, *Orothamnus zeyheri*. The *Elliottia* was encountered first in a limited area in Georgia in the valley of the Savannah River. It produces few viable seeds and vegetative propagation has proved difficult. The Orothamnus is confined to a very restricted area in South Africa. If the terminal flower of this unbranched stem is removed, there will not be any development of the bud and the plants die. The plants have a poor seed set and low germination. These factors have led the two species to lose their capacity of regeneration and adaptation and hence marching toward the path to extinction. There are a number of plant species that are sliding down the path to extinction for reasons that are directly or indirectly connected with the human activities.

There are a number of factors which have played a predominant role causing plants to be under threat. Visualising the colossal magnitude of the problem, the IUCN Threatened Plants Committee has given the following tentative list of threats:

1. Grazing
2. Regeneration of scrub (and lack of grazing)
3. Changes in arable farming
4. Ploughing of old grassland
5. Forestry
6. Traditional rural practices
7. Flooding
8. Drainage
9. Water pollution
10. Air pollution
11. Industrialization and urbanization
12. Road building
13. Tourism and tourist development
14. Dam construction
15. Mining and quarrying
16. Pressure from introduced plants
17. Collectors for horticulture
18. Collectors for botany
19. Critically low population—say below 100 individuals (hence IUCN category endangered)

20. Natural causes
21. Lack of pollinators
22. Other threats

Whereas Zeedyk et al. 1978 have suggested the following five main categories of threat :

1. The present threatened destruction or modification of habitats and ranges
2. Over-exploitation for commercial, scientific and educational purposes.
3. Diseases
4. Inadequacy of existing regulative mechanism
5. Other natural or man-made factors.

Ever since the beginning of civilization man began clear the land for agriculture and modernisation and this led him to push back natural vegetation and replace it by pastures and arable fields. This whole process has gone slowly for several centuries but recently manifold increase in population and the new technological developments have caused unbearable pressure on the earth. Uptil now the development of technology could keep pace with the population explosion in meeting the food requirements. This could be done by exploiting the potentialities of plants and animal species. For the fulfilment of basic requirements of our grandchildren, it is of paramount importance to arrest the decline in animal and plant populations. Even at the present trend it is expected that perhaps in a few decades there will be very little natural country left except in the most rugged areas, even in these there will be mines and wells and fertile land everywhere will be heavily cultivated. It is then that the importance of the national parks, botanic gardens and protected collections will be realised. One of our immediate tasks is to restore the ecosystems of the Himalayas and other mountain ranges. The genetic wealth is primarily confined in the tribal dominated areas which are in the wake of scientific and technical and industrial developments. Modern civilization coupled with the high yielding varieties has unequivocally threatened the existence or survival of native cultivars in these areas of cultivation. These local land races are probably the products of hundreds of years of natural and human selection. This genetic wealth possesses valuable characteristics like resistance to pests and diseases, adaptability to environment and quality, trade etc. The natural diversity for several crops prevails in forests i.e. for *Citrus*, *Musa* and *Mangifera* in the north-eastern hill regions. The Himalayas have considerably influenced the life, culture and history of this sub-continent. The slopes of the Himalayas are covered with enormous forests grasslands and other vegetation, including some of the most economic crop plants and beautiful flowering plants of the world. Several explorations in this region have discovered valuable plant wealth including the priceless gems like rhododendrons for the gardens of the world. These hills with the vast complex of valleys nourish agriculture, horticulture, forestry and animal husbandry in addition to the rich mineral resources. It is ironical that in a region of such enormous natural richness its dwellers are among the poorest in the country. It is envisaged that the development in these areas should proceed in such a way that the precious environment so important for this continent is not unnecessarily damaged and even efforts are made to preserve it in a suitable condition for the generations yet to come. Of the total area of nearly 11.0 lakh hectares under the different fruits in India over 1.0 lakh hectares is occupied by citrus fruits. Mandarin oranges, lime and lemon are grown extensively, whereas fruits like grape fruit are grown to a limited extent. There is considerable difficulty with the popular definitions of oranges and lemons because of the commonness in characteristics. The two types of existing sweet oranges, loose skinned and close skinned Portugal oranges have distinct characteristics. The loose skinned oranges have existed from a remote period in the eastern Himalayan region.

The total number of *Citrus* species in the world ranges between 16 and 157 but only 37 of those are collected in India. It has been stated unequivocally that the citrus fruits are native to Southern Asia. Hays (1966) stated that some citrus fruits originated from China and some from India while others might have originated somewhere between these two countries and in the Malay Archipelago. The north-eastern Himalayan region considered once to be a rich reservoir of genetic wealth of Citrus, has lost a considerable number of citrus fruits and the situation is alarming due to the socio-economic changes.

Every step taken by mankind towards progress has only increased man's dependence on nature. The advancement in civilisation coupled with increased population, has definitely threatened the life supporting systems. Forests do not serve the need only of the local communities but also people far beyond. Developing countries like India have suffered irrevocable damage by the destruction of innumerable diversified trees in the quest of food, fuel and fodder. The north-eastern Himalayan region comprising the states of Arunachal Pradesh, Assam, Manipur, Meghalaya, Mizoram, Nagaland, Tripura and Sikkim, is not an exception. This region is primarily inhabited by tribals practising the age-old methods of shift cultivation or 'Jhuming'. In this method large-scale clearing of forests is made during the winter months and burning of wood starts with the summer, for a year or two, then the yield of the crops slows down and new forest area are brought under this practice. The tribals are considered the custodians of the genetic wealth, since north-eastern region is primarily inhabited by them. The north-eastern Himalayan region has several pockets with enormous diversity e.g. *Citrus*, *Musa*, *Oryza*, *Saccharum*, *Mangifera* and *Zea mays*. The noteworthy among these is the prevailing enormous genetic diversity of *Citrus*. Bhattacharya and Dutta (1957) reported 17 *Citrus* species comprising 52 varieties in this region. The papedas, pumelos and their hybrids, *Citrus indica* and many types of citrons, lemons and mandarins are found in the eastern Himalayan region. It has been reported that *C. assamensis*, *C. indica*, *C. ichangensis*, *C. latipes*, *C. macroptera* are admittedly native to north-eastern region and possibly *C. megaloxycarpa*, *C. aurantinum* and one variety of *C. reticulata* are also indigenous.

The classification of citrus fruits is complex and this difficulty has increased tremendously by the numerous attempts made to classify the citrus fruits. Contradictorily these classifications instead of simplifying the problems have only made them more intricate. The change of climate, soil and culture, influence the features of the citrus fruits to some extent. The characters like smoothness, or wartiness, the degree of thickness of skin, the degree of sweetness, extent of juiciness and particularly the size of the fruit are likely to change. It is therefore suggested that a majority of the characters of the tree, flower, seed and fruit should be taken into account in grouping several types of citrus fruits. There is no classification which has found general acceptance. In principle, the original species must be described from the truly wild specimen. Botanists have deviated from this general rule, else they might have had to eliminate the entire *Citrus* genus, for even the generally accepted are found only in cultivation. The chances of difference between the cultivated and wild ones are more if these are farther removed in time. In the present studies, an attempt has been made to gather all possible available information uptil now and present it in a simple form. This information includes the available wild forms.

Recent investigations have provided more than 23 *Citrus* species, most of which are confined to the north-eastern region. Of the several species, *C. latipes*, *C. macroptera*, *C. ichangensis*, *C. assamensis* and *C. indica* were encountered in semi-wild and wild forms. These species are found growing in specific areas. It may prove disastrous if the citrus species with potential disease or pest resistance or the ability to withstand adverse conditions is not passed on to future generations. The old land races are increasingly replaced and gradually die out through neglect, as these are being used in farming, forestry, horticulture and other uses. It is economical to retain certain hybrids in addition to these land races which are of genetic value in breeding plants used in a variety of ways by human beings. Holding of living collections including rare and threatened or endangered species needs the maintenance of a proper recording system e.g. location of its prevalence, history and utility.

Citrus fruits are grown all over India. These occupy over 1.0 lakh hectares of which Andhra Pradesh, Maharashtra and Karnataka are the largest. The orange, lemon and lime are known for their taste and daily domestic use. The relative proportion of the sweet and sub-acid determines the taste of the orange. Limes and lemons make valuable drinks which serve to quench thirst, and counteract exhaustion in all types of people including sick persons. Almost all parts of citrus fruits are used in one or the other form. Delicious preparations like jelly, jam and marmalade etc. are prepared from it. These also provide nice perfumes and oils in additions to their valuable use as medicines.

The problem of collection and conservation of plants did not attract people early civilization due to small number of human beings. The present situation is alarming as enormous amount of destruction is being done due to one or the other reason. The conservation of plants cannot be postponed any longer. These collections can be made in seed, cutting, bulb or any other storage organ or in some cases it may necessitate the removal of the entire individual from its natural habitat. In other cases it may be essential to preserve the plant in its natural habitat. Systematic collection necessitates a knowledge of the precise site of genetic diversity. Alphonse de Candolle (1883) was the first scientist to study 249 species of cultivated plants and speculate on their origin employing four kinds of evidences history, philology or linguistics, botany with particular reference to geographic distribution of wild relations and archaeology. He was perhaps the first to indicate regions where plant domestication must have taken place. These are China, S-W Asia, Egypt and tropical Asia. N.I. Vavilov (1927) pointed out that the primary centre is noted by high frequency of the dominant alleles. The frequency of these alleles decreases towards the periphery and obviously the diversity decreases, thus these develop as secondary gene centres. The regions may overlap for cultivated plants. He suggested that the plant types within a group have some characteristics in common, which can conveniently be used as the guideline for distribution of a particular species. The number of cradles of agriculture has been discussed in depth by various workers. Although various types of classifications for centres of origin have been suggested the classification developed by Vavilov still forms the basis of subsequent work. The concept of Vavilov can be modified on the basis of more extensive information available on genetic diversity of crop plants. In one way or the other cultivated plants must have arisen from wild types but this has happened at very different points in time. Domestication brings manifold changes in plant character. Zeven and Zukovsky (1975) stated that citrus fruits have their origin in the Indo-Chinese-Indonesia centre, Chinese-Japanese centre, Mediterranean centre, Hindustani centre and also some in the African centre, South American centre, Australian centre and Central American and Mexican centre. The north-eastern Himalayan region is considered the natural home of many citrus fruits. Several primitive cultivars, wild types, related types and commercial types have been encountered in this region. Their variability may be carefully collected and conserved. For the systematic collection programme, the collectors should communicate well in time with the local concerned officials and should have a thorough knowledge of the various steps involved in collecting materials, appropriate methods of handling the collections and finally the various procedures involved in cataloguing and storage of collections.

It is expected that the explorer is well acquainted with the local cultivars, commercial types, wild types, primitive types and related species. It is particularly of interest in the case of citrus fruits having number of species each with numerous types. Some of the citrus species are still faced with controversy with regard to their systematic botany. These fruit crops are also faced with peculiar problems of being propagated through clones coupled with large size and longer life. The scientists are not in agreement with the precise specific position of some types among citrus fruits resulting in different names for the same types or viceversa. The precise estimate of the number of samples can be made at the time of travel while studying the existing variability.

Special attention should be focussed on remote areas to collect primitive varieties and related wild species of citrus fruits. The object of the collector should be to include as much variation as possible in a sample. Breeders require much larger diversity in agricultural plants than generally required for wild plants in maintaining a representative sample of population. It is stressed that while collecting wild species and related species, the old cultivars should also be collected. These are the important sources for resistance to pests and diseases. The explorer should take equal interest in primary and secondary centres of diversity. It is stressed that the periphery of their distribution should in no case be neglected as in such areas varied types with important traits like disease and pest resistance may be found. These geno types are adapted to conditions not found in the central area of distribution. Although in crops as highly cross pollinated as fruit crops it is desired to collect random samples, but due to practical limitation,it is advisable to resort to the system where random biased sampling is combined. Seeds are normally more convenient to collect and are preferred. Whenever the seeds are collected it is important to record the surrounding trees as potential pollen parents. Some of the characters are recorded at the site while others can be recorded in the laboratory. Each sample should be photographed; and the sample should be in the form of seed (fruit) or vegetative parts. If bud sticks and graft wood are collected, these should be properly packed in plastic covers and kept moist. These are relatively less bulky and convenient for transportation. These can be utilized by budding or grafting. The appropriate season for grafting is when the trees are dormant. Budding can profitably be done at the time when the sap is active in the root stocks. The desired root stock has to be arranged in advance. Citrus fruits can be propagated by seeds, budding, grafting, cutting or layering. The layering is not much in practice whereas budding by far is the most common method. Extensive studies have made it clear that the bulk of variability in cultivated plants is restricted to certain regions of the world and most of the regions are confined to developing countries. It is all the more ironical that in regions of much remarkable natural richness, its people are the poorest in the world. The developing countries possess the enormous genetic variability in economic plants and it is their responsibility to ensure that this useful variability is not lost due to the introduction of more productive varieties and land use patterns.

Living resources are renewable if conserved and are destructible if are not conserved. Rapid economic development without ecological damage can be reached only if conservation becomes a way of life with the common people. The conservation and development are mutually dependent. The process of development should be so designed as to conserve living resources. Conservation will be more effective if it also meets short-term economic needs. There are about fifteen hundred million people in the world who are dependent upon wood, dung, or crop waste for their fuel. Conservation is meaningless for rural communities if it does not meet their immediate requirements.

Dung and crop waste to an extent of four hundred million tons are burnt annually in different parts of the world. These could have been highly useful for regeneration of the soil. There are many such activities intimately associated with the very existence of mankind and are liable to bring changes in the ecosystem. A good example of this type is that brick-kilns are cropping up everywhere using fertile soil from the top layers. Brickmakers and brick users should be made aware that it takes nature over a century to make one centimetre of top soil. The haphazard digging of the soil for bricks could be prevented if brick making can be linked to the creation of permanent community assets such as farm pond or water-channels through a careful choice of sites for kilns based on the topography of a given area.

Man's thoughtless actions upsets the wonderful pattern evolved by nature. In this context, there are a few more noteworthy instances. The indiscriminate killing of snakes for their skins causes the increase of field rats. The re-emergence of malaria is due to the lack of an integrated approach to pest control.

The predator-prey relationship in wild life is of pivotal importance for keeping nature's balance. In Meghalaya, the reduction of a number of predators like the tiger and panther has resulted in the rapid increase of the population of wild pigs. Consequently, they have become a menace to the crop plants. The lack of sufficient co-ordination among those in charge of agriculture and fisheries leads to difficulties in sustaining ancient practices such as the rice-cum-fish or prawn culture. Development can be effective and more fruitful if it is guided by ecological, social cultural and ethical considerations. It is thus inevitable to ensure sustainable utilization of species and ecosystems and to preserve the range of genetic diversity of plants and animals.

The main ecosystem are also called life supporting systems. Agricultural, forests, coastal and fresh water systems are the threatened life supporting systems. There are about 1.50 lakh flowering plants of the tropical regions of the world. It seems that about two-thirds of the species of the plants of the world are of tropical distribution. Swaminathan (1979) mentions that narrow genetic base confined to highly specialised adaptations could lead to evolutionary blind alley and thereby to extinction. The haphazard interferencc of man in the ecosystems has increased the pace of extinction of species and of the genotypes within the species. IUCN has reported that an average of one species or sub-species of animal is lost each year and roughly 10% of the flowering plants of the world are led to under threat. About 4,500 of the total nearly 85,000 species for the entire temperate regions of the world could be considered threatened. Raven (1976) reported that according to estimates of IUCN Threatened Plants Committee (TPC), there are about 750 of these 4,500 species which should be considered endangered or vulnerable in the strict sense. It is indicated that within 25 years that no undisturbed tropical low land forest in the world will exist. It is believed that one-third of the tropical plant species will be heading towards extinction by the end of this century, with the first wave of extinction in the south-east Asia and New Guinea taking place within the next decade. The formidable task of preservation of genetic wealth is of paramount importance for improving agricultural, forestry and fisheries production. The usefulness of these materials for future options as a raw material for scientific and industrial innovations is of utmost importance. It is beyond our reach to predict that what genetic materials may become useful. Frankle (1973) mentions that the alarming situation caused by the narrow genetic base in crops such as wheat, rice, millets, beans, banana, lime and oranges could be taken care of by gathering the range of existing genetic variation in the wild and domestic types.

There are broadly two ways of protection of genetic variability:

1. (i) **Off Site** : Part of the organism: This involves the preservation of seed, semen or other element from which the organism concerned can be reproduced.

(ii) **Off Site (ex citu)**—Whole organism: The stalk of the individual of the organism concerned is kept out of its natural habitat in plantation, botanical garden, zoo, aquarium and ranch or culture collection.

2. **On Site (in situ)** : The stalk is preserved by protecting the ecosystem in which it occurs naturally.

1. (i) **Off Site** : It is essential to establish long-term seed storage facility to serve as base collection. The usefulness of the gene bank lies in whether these are fed with the proper germplasm. Conservation is not just seed storage. The germplasm should be collected first and then be evaluated. The seed collecting will act as assurance against the complete destruction in the field in cases where only a few scattered strands of an extremely rare species are known. Reconstruction of the ecosystem may be resorted to at later stages using the seed bank material. This may not be necessary if adequate biosphere reserves are established. The active co-operation of different countries is urged for the maintenance of collections requiring specific environments. There is mounting concern for establishing a virus free site for the *Citrus* material to be collected in Sri Lanka. The following problems are intimately associated with the conservation of genetic materials :

1. Long-term storage,
2. Genetic integrity of stored seed,
3. Regeneration of genetic materials,
4. Documentation and exchange of information, and
5. Quarantine problems.

(ii) **Off Site (ex situ)** : The botanic gardens find themselves a purposeful role in developing living collections for their manifold potential functions in conservation. These collections play a prominent role as an important source of conservation oriented research and education. The botanic gardens also play a major role in conveying information about conservations to the general public. The sympathy and understanding of the millions of people for conservation is improved tremendously through these gardens. These can also be used as a source for re-introduction into the wild. There are many species whose future is unlikely in cultivation and the task of maintaining them will be a sentimental burden which is not justifiable by any standard. In well-laid-out botanic Gardens, certain important ideas can be practically presented to the public by careful planting and discreet levelling :

1. The importance of plants to human life,

2. The practical usefulness of conservation, and
3. The suitability of the range of plant life for small domestic gardens.

These botanical gardens with deep and far reaching impact should aim at the following :

1. Assemble speciality,
2. Proper record of all items,
3. Preservation of collections,
4. Increase them, and
5. Disperse them as it is the surest way of preserving them as there is a saying "if you want to keep a plant, give it away".

It is not possible for any single organisation to tackle the problem of conservation. The role of international organisations is vital. Sometimes it becomes the duty of these in-charge of living collections to make the gardeners aware about the rare species. It can be exemplified by the two of the oldest mutations, the laciniate form of *Chelidonium magus* and the entire leaved form of *Urtica pilulifera* which have no decorative value but have great evolutionary and historical significances. It is felt that the life of cultivated species cannot be extended indefinitely partially because different breeding methods may be used in future. It is, therefore, stressed that the following categories of plants should be preserved :

1. The plants of historical value in the evolution of horticultural and agricultural plants.
2. Plants with known genetic value.
3. Plants with known characteristic traits e.g. resistance to pests and diseases.
4. Preserving a gene reservoir for basic and applied research and education. Education is a function of botanic gardens required to gain scientific and public interest.
5. Conservation of rare plants.
6. Conservation of plants whose economic potential has not yet been investigated.

It is essential in the process of conservation of domesticated plants that they should be duplicated. The seed and plant material may be shared with different organisations.

The conservation oriented living collections should be envisaged as repositories of rare and diminishing species. There is a need for a better flow of information regarding the technology of living collections like growing techniques, propagation methods etc. The well organised collections can be used to re-establish endangered plant species in their area once the danger that caused their depletion has been overcome. There is the need for the following information for an integrated international policy :

1. The botanic garden willing to take part in international policies should furnish a detailed account of the contents of its living collections. The Indian neem tree so well-known for its several uses in sickness and health has now sadly been neglected in our country. At the same time it is being planted for greenery in what was once a desert area of Nigeria. The exchange of information of such types will be of abundant use.
   The American Horticultural Society's Plant Sciences Data Centre has made a significant contribution by establishing a centralized computerized inventory of cultivated plants within protected collections such as those in botanical gardens, arboreta and parks of North America. This information bank includes the accession records of about 26 botanical gardens including the records from Fairchild Tropical Garden (Florida), Longwood Gardens (Delaware), Morton Arboretum (Illinois), United States National Arboreta (Washington DC), Los Angeles State and Country Arboreta and Botanic Garden (California) and the Royal Botanic Garden (Ontario, Canada).
2. The aim of domestic policies of individual garden i.e. financial social and educational.
3. The notification of likely national and international expeditions.
4. The dissemination of information on the techniques of

the work in botanic gardens e.g. growing techniques, propagation and record systems.

Seemingly, many of the established botanic gardens are crowded with plants and grow a similar range of plants. This could be avoided if there is a proper understanding and a collaboration between national and international organisations dealing with conservation. The possibility of growing many trees and vigorous shrubs that would safely be grown outside of botanic gardens should be explored as these would leave room for species whose maintenance and cultural requirement are more exciting. Many factors operate in the selection of the site of the garden; the following seem to be most important :

1. The climate of the site and the provision of controlled environment.
2. The availability of efficient staff.
3. The nature of the available material : from expeditions, exchange between conservationists or through trade.
4. The purpose served by the garden : amenity, research, education and conservation.

Private gardens have played a vital role in the preservation of rare plants, but often not deliberately. The varied tastes apart from different climates and local conditions throughout the world makes the preservation of rare types essential. It is stressed that tropical rain forest is one of the most highly endangered habitats and obviously this necessitates the need for botanic gardens in the tropics for preservation of the tremendous range of the available genetic wealth.

2. **On Site (in situ)** : Sometimes the biotypes of rare species are difficult and costly to conserve and thus it will be more profitable if the public is made aware of the necessity of protecting small areas of the natural habitat of such plants. The people can easily be convinced if the plants are of some economic use to them.

It is not possible to restore the whole spectrum of ecological conditions in botanic gardens since generally they deal with the small isolated populations of single species and not with the biotypes. The genotypes are exposed to continuous alternation by a recombination of genetic material due to the sexual reproduction. The changed environmental conditions result in the selection out of control and consequently will follow different lines from those in nature. *In situ* conservation allows the populations to follow their natural evolution and thus it is possible to maintain the range of existing variability. It is visualized that perhaps in the near future, there will be little natural country left, except in the most rugged areas, and fertile land everywhere as pointed out earlier, will be heavily cultivated. It is then that national parks will come into their own, obviously there is a pressing need for establishing national parks.

The varied number of staple foods, timbers, medicines, ornamental plants and other essential products would not have been available to us today if the plants from which they are derived or have been bred had not originally been collected. The problem of conservation did not seem to be so pressing in the old days when nature seemed so infinitely beautiful and the number of human beings was relatively small. To embellish our gardens, parks and public plantings is perhaps the main problem of the present day situation. Huxley (1974) mentioned that today, plant species with varied number from 50 to 200 become extinct annually, presumably the proportion will gradually increase. These will include plants for which man will become poorer. Immediate attention needs to be paid now else next week will be too late for at least one plant somewhere. Considering the large number, the priorities at the national and international level have to be fixed for this formidable task of conservation. It is generally considered that species endangered throughout their range and the species that are sole representatives of their family are species which deserve priority. It is considered that the greater the genetic loss, the less imminent that loss needs to be to justify preventive action. The extent of potential genetic loss is related to taxonomic hierarchy. Different positions in this hierarchy tend to reflect genetic differences in such variables as morphology, physiology, chemistry and ecology. The extent of difference between genera and between species within genera varies. The current taxonomic hierarchy is the only means for providing for the degree of potential loss of genetic material. The monotypic families and genera should receive priority over polytypic ones. The smaller the size of family the greater the magnitude of the differences as expected between the nearest related family; and similarly one group of species will be more distinct from the species of the other group. It is, therefore, suggested that an endangered species should be given priority over a vulnerable, vulnerable over a rare species, and rare species over one insufficiently threatened to qualify for one of the three types. This has already been stressed earlier. The precise extent of threat is a matter of state of knowledge but with highly restricted distribution e.g. *Citrus indica* should be immediately protected. The species with highly restricted distribution needs a close watch and particular attention should be paid to families and genera. A set of conservation categories of threat which are parallel to those used for animals are as follows :

1. Endangered : in danger of extinction. Survival unlikely without protective measures.
2. Vulnerable : not yet endangered but likely to be endangered if the causal factor continues operating.
3. Rare : threat of extinction but not yet endangered or vulnerable, world population is small in specialized habitats, so that it would be possible to endanger them or even make extinct e.g. by a sudden change in land use in one area.

Although the uncommon and inaccessible plants may or may not be faced with the danger of extinction they deserve to be preserved because they have potential food, medicinal or some other value.

The natural diversity for many crops prevails in forests i.e. for *Citrus*, *Musa* and *Mangifera* in the north-eastern Himalayan region. It was envisaged to preserve on a national level the important material by way of gene sanctuaries. The gene sanctuary is a mechanism to preserve the genetic diversity of endangered species by protecting the ecosystem in which it occurs naturally. Hence it was proposed to preserve the pockets of habitat where considerable genetic variability in the endangered species of *Citrus*, e.g. *Citrus indica* occurs. It was decided to establish a gene sanctuary—National Park for *C. indica* in Meghalaya within the overall concept of biosphere reserves. The biosphere reserves in general should have the following features (Anonymous, 1977) :

1. biosphere reserve will be protected areas of land and coastal environment;
2. the network of biosphere reserves should include significant biomes and their sub-divisions throughout the world;
3. each biosphere reserve will constitute one or more of the following categories :

   (i) representative examples of natural biomes;

   (ii) areas with unusual natural features or unique communities of exceptional interest. It is recognized that representative areas may constitute unique features like,

      (a) one population of a globally rare species, and

      (b) their representative characters and uniqueness may both be characteristics of an area;

   (iii) areas of harmonious landscape resulting from traditional patterns of land-use; and

   (iv) areas of modified or degraded ecosystem capable of being restored to more natural conditions;

4. each biosphere reserve should be large enough to be effective conservation unit and to accommodate different uses with ease;
5. they should provide opportunity for ecological, research, education and training. Their existence may be useful to other projects in the Man and Biosphere (MAB) programmes;
6. a biosphere reserve must have long-term legal protection;
7. in some cases biosphere reserve may coincide with or incorporate existing or proposed protected areas, such as national parks, sanctuaries or nature reserves.

It is quite predictable that the concept of biosphere is sufficiently flexible and will continue to evolve with time. The very concept of gene sanctuary was responsible in conceiving the establishment of a *Citrus* Gene Sanctuary—National Park. The recent studies have supported, on all counts, the earlier work that the north-eastern Himalayan region is the natural home of many *Citrus* species. It has been unquestionably admitted that *Citrus indica*, *C. assamensis*, *C. latipes*, *C. ichangensis* and *C. macroptera* are natives of north-eastern Himalayan region. These were observed in wild and semi-wild form. The first gene sanctuary in India for *Citrus* in Garo Hills within the concept of biosphere reserve as National Park will be established. The number of steps followed to develop this concept for the establishment of *Citrus* Gene Sanctuary are as follows :

1. The prevalence of a number of *Citrus* species including its wild species in the time scale in different parts of the country was studied by intensive personal surveys, personal interviews with the local tribals, officials and included the in-depth study of national herbaria of Botanical Survey of India at Howrah, Shillong and Dehra Dun and Forest Research Institute, Dehra Dun. The in-depth study of the history with particular reference to botany and philology was made pertaining to origin, classification, commercial uses and its association with different regional dwellers and centres of diversity. The natural diversity of *Citrus indica* was observed in the forests of the Garo Hills in Meghalaya. Plant characters varying from bush to the extent of climber were encountered. In order to allow the populations to follow the same line of natural evolution, on site preservation in the natural environment is essential. The frequency distribution of this species was recorded. The plants in the dense forest were relatively more prolific and were found in flowering stage than those at the periphery. The *C. indica* is considered as the most primitive and perhaps the progenitor of the *Citrus* species. The priority of preservation is essentially based on the degree of its being faced with the danger of extinction. It is an endangered

species and is confined to certain pockets in the forest of its native place. It possesses a number of traits like resistance to diseases and pests, adaptability to wider range and several other dasirable characters of economic importance which could be profitably utilized.

2. The wild orange (C. indica) has been encountered in the unique area with unusual features of exceptional interest. A fragment of Tura range in Garo Hills in Meghalaya is the most ideally suitable site for the establishment of *in situ Citrus* Gene Sanctuary—National Park. Both representativeness and uniqueness are the characteristics of this area.
3. A committee of distinguished scientists headed by the eminent scientist Dr. B.P. Pal, F.R.S., Chairman National Committee on Environmental Planning and Co-ordination was constituted to establish the first gene sanctuary in India for *Citrus* in Garo Hills. On the recommendations of the Committee, Government of Meghalaya was requested to declare the site as National Park under section 35 of the Wild Life (Protection) Act, 1972. The legislation concerning living resources is being strengthened in India. The laws which are ecologically, economically, and socially viable can only be enforced in such backward areas.
4. This National Park will open up the grounds for research, education and training. Ecological researches will be useful for providing benchmarks or standards for long-term changes in the ecosystem as a whole. The core area of the site is over 10265.96 hectares. It is undisturbed and large enough to be an effective conservation unit for *C. indica* and accommodating different uses without conflict. Possibilities are being explored to extend this area and to protect the fauna of this ecosystem. In addition to core zone, it will have a demarcated buffer zone. This will provide opportunities for co-ordinated research where agencies like the Indian Council of Agricultural Research; Department of Environment; Department of Science and Technology; Botanical Survey of India; Zoological Survey of India; Forest Department, Government of Meghalaya; and various

other departments work together. The research activities can conveniently be grouped as continuous or frequent, periodical and occasional. The programme pertaining to a specific nature should be developed for effective training at the post-graduate levels including research fellows and training courses and exchange of persons. The involvement of local communities is a valuable means for the successful achievement of these targets. Their participation in planning, decision making and management is indispensable.

5. The proposed site of the National Park will be an unique ecosystem which may later coincide with or incorporate with the bio-sphere reserve.

Bibliography

Anonymous, 1974. Trends, priorities and needs in systematic evolutionary biology. *Systematic Zoology*, 23: 416-29.

Anonymous, 1975. Preservation plots in India. *Indian Forest Bulletin* No. 271 (new series). *Silvics*, pp. 42, Publ. F.R.I. Press, Publicity and Liaison Branch, Forest Research Institute and Colleges, Dehra Dun.

Anonymous, 1976. Report of National Commission on Agriculture, Part IX on Forestry.

Anonymous, 1977. *Indian Agricultural Statistics* 1967-68 to 1969-70, Vol. II. Directorate of Economics & Statistics, Ministry of Agriculture & Irrigation, Government of India.

Anonymous, 1977. Draft Report MAB Report Series No. 45, pp.1-79, Scientific workshop on biosphere reserves in the Mediterranean Region—Development of a conceptual basis and a plan for the establishment of the regional network—Unesco side, Turkey, 6-11, June 1977.

Anonymous, 1978. Report IBPGR/Government of India, South Asian Plant Genetic Resources Workshop.

Anonymous, 1980. *World Conservation Strategy*. Document prepared by the International Union for Conservation of Nature and Natural Resources (IUCN) with the advice, co-operation and financial assistance of UNEP and WWF in collaboration with FAO and Unesco.

Babar, Z.D.M. 1519. *Memoirs of Zeher-ud-Din Muhammed Babar* (English Translation by Dr. John Leyden and W. Erskine 1826).

Basu, B.C. 1906. Orange cultivation in the Khasi Hills, *Agri. Sci.* 1 (I).

Bene, J.G., Beall, H.W. and A. Cote. 1977. *Trees, Food and People*, IDRC, Ottawa.

Bhattacharya, S.C. and Dutta, S. 1951. Citrus varieties in Assam. *Indian J. Genet. & Pl. Breed*, 2(I): 57-62.

Bhattacharya, S.C. and Dutta, S. 1956. Classification of citrus fruits of Assam. Publ. ICAR, Scientific Monograph, No. 20.

Bonavia, E. 1890. *The Cultivated Oranges and Lemons etc. of India and Ceylon*. Publ. W.H. Allen & Co., London (pp. xiv+384+Plates CCLIX).

Borthakur, D.N.; Prasad, R.N.; Ghosh, S.P.; Singh, A.; Awasthi, R.P.; Rai, R.N.; Verma, A.; Dutta, H.H.; Sachan, J.N. and Singh, M.D. Agro-forestry based forming system as an alternative to Jhuming. Paper presented at National Seminar on Agro-forestry, Imphal, May 16-18, 1979.

Brandis, D. 1874. *The Forest Flora of North-West and Central India*, pp. 50-6

Budowski, G. 1976. The global problems of conservation and the potential role of living collections. pp. 9-13 (in) *Conservations of Threatened Plants* (ed.) by Simmons, J.B.; Beyer, R.I.; Brandham, P.E.; Lucas, G. Li; and Perry, V.T.H. Publ. Plenum Press, New York, London.

Candolle de A. 1886/1912. *L'Origine desplantes cultive's* (*Origin of Cultivated Plants*) 5th edition, Paris.

Chatterjee, D. 1948. Botanical nomenclature of Shaddock or Pomelo. *Nature* 161, 770.

Conklin, H.C. 1957. Hanunoo agriculture—A report on an integrated system of shifting cultivation in the Philippine, FAO forestry development Paper No. 12 FAO, Rome.

Darlington, C.D. 1956. *Chromosome Botany and the Origins of Cultivated Plants*. Revised 2nd ed. G. Allen and Unwin Ltd. pp. 231.

Darlington, C.D. and E.K. Janaki Ammal. 1945. *Chromosome Atlas of Cultivated Plants*. G. Allen & Unwin, London, pp. 397.

Dutta, S. 1958. Origin and history of citrus fruits of Assam. *Indian Hort*. 15 (3 & 4): 147-53.

Dutta, S. and Bhattacharya, S.C. 1951. A new citrus species of Assam. *Indian J. Hort*.

Frankel, O.H. 1973. Survey of crop resources in their centres of diversity: FAO/IBP.

Gandhi, Indira. 1980. Key Note Address—Launch of the world conservation strategy in India, Vigyan Bhavan, New Delhi, March 6 1980, p. 6.

Ghosh S.P. 1977. Citrus industry of north-east India. *Punjab Hort., J.* 17 (1 & 2): 13-21.

Glidden, H.W. 1937. The lemon in Asia and Europe. *Amer. Oriental Soc. J.* 57, 381-96.

Harlan, J.R. 1971. Agricultural origins : Centres and non-centres, *science*. N.Y. 174: 468-74.

Harlan, Jack, R. 1972. Genetic disaster. *Journal of Environmental Quality* 1: 212-15.

Hayes, W.B. 1960. The citrus fruits. pp. 196-266 (in) *Fruit Growing in India* (3rd Revised Edition). Publ. Kitabistan, Allahabad, pp. 512+10.

Hooker, J.D. 1872. *The Flora of British India*, I, pp. 514-16.

Hume, H. Horold. 1966. *Citrus Fruits*. Publ. The Macmillan Co., New York. Collier-Macmillan, Canada Ltd. Toronto Ontario.

Huxley, A. 1974. The ethics of plant collecting. *Journal of the Royal Hort. Soc.* 99 (6): 242-49.

Jain, S.K. 1981. *Plant Conservation Bulletin No. 1*, pp. 13. Botanical Survey of India, Howrah-711103

Kanjilal, P.C.; Das, A. and Purkayastha, C. 1936. *Flora of Assam* I (II), 212-15.

King, F. Wayne. 1978. The wild life trade (in): Howrad, P. Brokaw (ed.) *Wildlife in America*. Council on Environmental Quality, US Fish and Wild Life Service, National Oceanic and Atmospheric Administration, Washington D.C.

King, K.F.S. 1979. Some principles of Agroforestry. Paper presented at the National Seminar on Agroforestry, Imphal, p. 18, May 16-18, 1979.

King K.F.S. and Chandler, M.T. 1978. The Wasted Lands, ICRAF, Nairobi, p. 35.

Kruz, S. 1877. *Forest Flora of British Burma*. pp. 184-89.

Laufer, B. 1934. The lemon in China and elsewhere. *Amer. Orient. Soc. J.*, 54 : 143-60.

Lindley, J. and Moore, T. 1866. Treasury of Botany, (1) 291.

Lowther, Capt. 1857. The flora and vegetable products of Assam, *Jour. A.H.S. of India.*

Lushington, A.W. 1910. The genus citrus, *Indian Forester*, 36: 323-53.

Melville, R. 1970. Plant conservation in relation to horticulture. *Journal of the Royal Horticultural Society* 95 (11): 473-80.

Michael, A.M.; Sarma, P.B.S.; Ashok Raj, P.C. and Chawdhury, Ram Kumar. 1978. *Resources Inventory Garo Hills* (Meghalaya) Publ. ICAR, New Delhi. p. 320.

Montrouzier. 1860. Flora de I'lle Art in Mem. Acad. Cyon. Coll. Sci. 10187 (Original not seen, quoted from *Cit. Indry.* 1437).

Mitra, S.K. and Khongwir, P.C. 1928. Orange cultivation in Assam. Dept. Agr. Assam. Bull. No. 2.

Naik, K.C. 1940. A study of the pre-orchard life of certain rootstocks for Chinese orange and acid lime at Kodur, *Indian. J. Agric. Sci*. 10 (4), 602.

Playfair, Major, A. 1975. *The Garos*. Publ. United Publishers, Pan Bazar, Gauhati, pp. xxvi+172.

Raven, H. Peter 1976: Ethics and attitudes, pp. 155-79 (in) *Conservation of Threatened Plants* (ed.) by Simmons, J.B.; Beyer, R.I.; Brandham, P.E.; Lucas, G. Li and Parry V.T.H. Publ. Plenum Press, New York.

Risso, J.A. 1813. Essai sur I'historie des oranges bigaradiers, limettiere, cedratiers, limoniers ou Citronniers Cultives dansle Department de desalpes Maritimes. Dufour, Paris, 74, 2 pls.

Risso, J.A. and A. Poiteau. 1818. Histoire naturelle des orangers. Audot. Paris. 280.

Robinson, W. 1841. *Account of Assam.*

Roxburgh, W. 1832, 1874. *Flora Indica.*

Roylc, J.F. 1839. *The Illustrations of Botany of the Himalayan Mountains Wm.* II. Allcn & Co., London.

Saucer, C.O. 1952. *Agricultural Origins and Dispersals*, pp. 175. The M.I.T. Press, New York.

Schlippe, P.D. 1956. *Shifting Cultivation in Africa*, Routledge and Kegon Paul, London.

Singh, Bhag 1977. *Races of Maize in India*. Indian Council of Agricultural Research, p. 106.

Singh, Bhag 1978. *In situ* preservation of variability of Citrus species *Indian. J. Genet. Pl. Breed.* In press.

Singh, Bhag 1979. Setting of the *in situ Citrus* Genc Sanctuary in Meghalaya. Paper presented at "National Seminar on Agro-Forestry", Imphal, May 16-18, 1979. (In press)

Singh, Bhag 1980. Establishment of First Gene Sanctuary in India for *Citrus* in Garo Hills. Paper presented at "Launch of world conservation strategy", Vigyan Bhavan, New Delhi, March 6, 1980.

Singh, Bhag 1980. *On site* preservation of *Citrus* in Meghalaya. Paper presented at symposium on "Development without Destruction", Shillong, October 3-4, 1980. (In press).

Singh, Bhag 1980. Garo Gene Sanctuary of *Citrus Voluntary Action*, XXII (5) : 27.

Singh, Bhag 1981. *In situ* Conservation of genetic resources of eastern Himalayan region with special reference of citrus paper presented at the seminar on resource potentials of north-east India, 19th April—1st May 1981 held at State Central Library, Shillong. (In press)

Singh, Harbhajan 1957. Genetic stocks of agricultural and horticultural plants of South Asia. International symposium on Genetics and Pl. Breed. in South Asia. *Indian. J. Genet Pl. Breed.* 17 (2): 343-61.

Singh, H.B. and Arora, R.K. 1978. *Wild Edible Plants of India.* pp.vi + 88, Publ. ICAR, New Delhi.

Singh, Ranjit 1977. Indigenous *Citrus* germplasm in India, pp. 39-49 (Ed. Nijjar, G.S., *Fruit Growing in India*). Oxford and IBH Pub. Co., New Delhi.

Singh, R.N. and Sharma, D.K. (Unpublished). Plant genetic resources and their utilization in fruit crops. National symposium of plant and animal genetic resources, Delhi, 28-30th Dec. 1978.

Stracey, P.D. 1966. Wild life management in India. Publ. Leaflet No. 3, Indian Board for Wild Life. Min. of Food and Agric., Department of Agriculture, New Delhi.

Swaminathan, M.S. 1978a. Inaugural Address-National Symposium on plant and animal genetic resources, New Delhi, 28-30th. Dec. 1978, pp. 20.

Swaminathan, M.S. 1978b. Inaugural Address—Indian Botanical Society, First Botanical Conference, Meerut. Dec. 20-30th, 1978, p. 31.

Swaminathan, M.S. 1980. Keynote Address—FAO/UNEP Technical consultation on animal genetic resources conservation and management, Rome, 2-6th June, 1980, p. 10.

Swingle, W.T. 1913a. Botanical name of the lime (*Citrus aurantifolia*) *Wash. Acad. Sci. J.* 3, 463-65.

Swingle, W.T. 1913b. *Citrus ichangensis*, a promising hardy new species from *South Western China and Assam. J. Agri. Res.* I, 1-4 (Original not seen).

Swingle, W.T. 1943. The botany of *Citrus* and its wild relatives of the orange sub-family (family Rutaceae, sub-family Aurantioideae) Chapter 4 of the *Citrus* Industry, Vol. I, Univ. of Calif. Press, Berkeley and Los Angeles.

Swingle, W.T. 1946. Botany of *Citrus*, Chapter IV in the *Citrus* Industry I, pp. 129-474 Univ. California Press, Berkeley.

Sykes, J.T. 1975. Tree crops. In *crop Genetic Resources for Today and Tomorrow* (ed. O.H. Frankel and J.G. Hawkes) IBPI, pp. 123-37. Cambridge Univ. Press, Great Britain.

Tanaka, T. 1928. On certain new species of *Citrus. Studia Citrologica*, 2:155-64.

Tanaka, T. 1937a. Citrus fruits of India. Allahabad Farmer, 11 : 2.

Tanaka, T. 1937b. Further revision of Ruteceae—*Aurantio-ideae of India and Ceylon. J. Ind. Bot. Soc.* 16 : 227-40.

Tanaka, T. 1937c. Origin and development of citrus species and varieties. *Communication.* Hort. Inst. Taihoku Imp. Univ. (Japan) No. 66.

Tanaka, T. 1954. Species problem in *Citrus.* Japanese Society for the promotion of Science. Ueno. Tokyo.

Thomas, G.S. 1976. The National Trust pp. 35-9 (in) *Conservation of Threatened Plants* (ed.) by Simmons J.B ; Beyer, R.I.; Brandham, P.E.; Lucas, G.Li and Parry, V.T.H. Publ. Plenum Press, New York and London.

U.S. Rep. 1967. *The World Food Problem*, Vol. II, Washington.

Vavilov, N.I. 1926. Studies on the origin of cultivated plants, Bull. Appl. Bot. 16(2) : 248.

Vavilov, N.I. 1928. Geographic Genzentren unserer Kulturpflanzen. Int. Kongr. G. Vererb. Wiss (1927). Z. indukt. Abstammu—Vererblehre Suppl., 1-34.

Vavilov, N.I. 1940. The new systematics of cultivated plants pp. 519-66 (in) Huxley (ed.), *The New Systematics*, Oxford University Press.

Walters, S.M. 1973. The role of botanic gardens in conservation. *Journal of the Royal Horticultural Society* 98 (7) : 311-15.

Watt, G. 1889. *A Dictionary of the Economic Products of India.* (reprint ed.), Cosmo Publ. Delhi, India.

Watt, H. 1908. *The Commercial Products of India.* John Murray, London.

Webber, H.J. 1946. Cultivated varieties of Citrus, Chapter V, in *C trus* Industry, I, 475-668.

Webber, H.J. and Bachelor, L.D. 1946. The *Citrus* Industry I, University of California Press, Barkeley.

Wight, R. 1940-50. *Illustration of India Botany.* Pharoh Madras (Original not seen).

Zeedyk, W.D.; Farmer, R.E.; Mac Bryde; Jt. B. and Baker, G.S. 1978. *Endangered Plant Species and Wildland Management.* J. For. Jan. 31-8.

Zeven, A.C. and Zhukovsky, P.M. 1975. *Dictionary of Cultivated Plants and their Centre of Diversity*, excluding ornamentals, forest trees and lower plants. p. 219, Publ. Centre for Agricultural Publishing and Documentation, Wageningen.

Zhukovsky, P.M. 1968. (New Centres of origin and new gene centres of cultivated plants including specifically endemic microcentres of species closely allied to cultivated species). Bot. Zh. 53 : 430-60 pl. Breed. Abstr. 38(38).

Index